Pan Study Aids
===

Chemistry

Mike Denial

Pan Books London and Sydney
in association with **Heinemann Educational Books**

First published in 1980 by Pan Books Ltd,
Cavaye Place, London SW10 9PG
in association with Heinemann Educational Books Ltd
© Mike Denial 1980
ISBN 0 330 26080 4
Printed in Great Britain by
Richard Clay (The Chaucer Press) Ltd, Bungay, Suffolk

This book is sold subject to the condition that it
shall not, by way of trade or otherwise, be lent, re-sold,
hired out or otherwise circulated without the publisher's prior
consent in any form of binding or cover other than that in which
it is published and without a similar condition including this
condition being imposed on the subsequent purchaser

PAN STUDY AIDS

Titles published in this series

Biology
Chemistry
English Language
Essential Study Skills
Geography 1.   *Physical and Human*
Geography 2.   *British Isles, Western Europe, North America*
Maths
Physics

Forthcoming titles

Book-keeping and Accounts
Commerce
Economics
French
History 1.   *British*
History 2.   *European*
Human Biology

Brodies Notes on English Literature

This long established series published in Pan Study Aids now contains more than 150 titles. Each volume covers one of the major works of English literature regularly set for examinations.

Contents

Acknowledgements 6
Using the book 6

Unit 1 **Atomic structure, kinetic theory and bonding**
    1.1 Atoms, molecules and ions   9
    1.2 Atomic structure   11
    1.3 The kinetic theory   14
    1.4 Symbols, formulae and bonding. Molecular shapes   21
    Questions   33

Unit 2 **Elements, compounds and mixtures. Purification**
    2.1 Elements, compounds and mixtures   36
    2.2 Purification techniques   39
    2.3 Criteria of purity   45
    Questions   47

Unit 3 **Electrochemistry. The electrochemical series, oxidation and reduction**
    3.1 The conduction of electricity by chemicals   49
    3.2 The electrochemical series (activity series)   56
    3.3 Oxidation and reduction   62
    Questions   65

Unit 4 **Acids, bases and salts**
    4.1 Acids and bases   67
    4.2 pH and neutralization. Salts   71
    4.3 Strong and weak acids and bases   76

Unit 5 **The air, gases, oxygen and hydrogen**
    5.1 The air   78
    5.2 Combustion, respiration, photosynthesis and air pollution   79
    5.3 Gas preparations   82
    5.4 Oxygen   84
    5.5 Hydrogen   86
    Questions   87

4 Contents

Unit 6 **The Periodic Table. Structure**
    6.1 The Periodic Table   89
    6.2 Structure   94
    Questions   99

Unit 7 **The mole. Chemical formulae and equations**
    7.1 The mole   100
    7.2 Percentage composition. Formulae   103
    7.3 Equations   107
    Questions   111

Unit 8 **The metals and their compounds**
    8.1 Metallic oxides   114
    8.2 Metallic hydroxides   117
    8.3 Metallic carbonates   118
    8.4 Metallic nitrates   120
    8.5 Metallic sulphates   121
    8.6 Metallic chlorides   123
    8.7 Sodium and potassium compounds   124
    8.8 Calcium, magnesium and zinc compounds   126
    8.9 Aluminium compounds   127
    8.10 Iron and its compounds   128
    8.11 Lead, copper and their compounds   130
    Questions   131

Unit 9 **Water, soaps and detergents. Hydrogen peroxide**
    9.1 Water as a chemical. Water pollution   133
    9.2 Solutions. Solubility curves   135
    9.3 Soaps and synthetic detergents   137
    9.4 Hardness of water   139
    9.5 Hydrogen peroxide, H_2O_2 141
    Questions   143

Unit 10 **More non-metals**
    10.1 Carbon and silicon   145
    10.2 Sulphur and its compounds   149
    10.3 Nitrogen and phosphorus   154
    10.4 Chlorine and the halogens   161
    Questions   164

Contents 5

Unit 11	*Analysis*	
	11.1 Volumetric analysis	166
	11.2 Tests for common gases and ions	169
	Questions	171

Unit 12	*Organic chemistry*	
	12.1 Hydrocarbons	172
	12.2 The oil industry	177
	12.3 Alcohols, acids and esters	178
	12.4 Polymers	181
	Questions	185

Unit 13	*Energy in chemistry*	
	13.1 Energy changes	187
	13.2 Fuels	191
	Questions	194

Unit 14	*Rates of reaction. Equilibria*	
	14.1 Rates of reaction	195
	14.2 Reversible reactions and equilibria	199
	Questions	203

Unit 15	*The chemical industry*	
	15.1 Social and economic factors	205
	15.2 Common industrial processes	205
	Questions	211

Unit 16	*More calculations*	
	16.1 Faraday's laws of electrolysis	212
	16.2 Volume changes in gases	214
	Questions	217

Unit 17	*Preparing for the examination*	
	17.1 Revising for the examination	218
	17.2 Answering multiple choice questions	220
	17.3 Answering structured questions	221
	17.4 Answering longer questions	222

Answers to questions   230

Periodic Table   232

Index   234

Acknowledgements

I would like to thank my colleague, Chris Knowles, for his reading of the initial draft of the book, and for his many detailed and helpful comments. Similarly, my thanks are due to Martyn Berry for his reading of the typescript, and for his many constructive comments.

Some of the questions at the end of each unit are reproduced by kind permission of the following Examining Boards, whose addresses are given on page 8:

The Associated Examining Board, University of Cambridge Local Examinations Syndicate, The Joint Matriculation Board, Oxford and Cambridge Schools Examination Board, The Welsh Joint Education Committee

Mike Denial, 1980

Using this book

This book will assist anyone faced with an 'O' level examination in chemistry, physical science or physics-with-chemistry, irrespective of the particular syllabus being followed. It can be used during the teaching programme to reinforce the principles being taught, and as a revision book in preparation for the examination. In this latter sense it provides more than a survey of all of the important facts and principles; it is a complete guide on how to prepare for an examination in 'O' level chemistry.

Using this book 7

All of the necessary facts have been depicted in clear and simple form. The main themes which are fundamental to an understanding of modern chemistry courses have been thoroughly discussed and correlated, and wherever possible these have been summarized in simple form for revision purposes. Common misunderstandings and examination mistakes have been stressed where appropriate, and advice given on answering the different types of examination question in current use. Model answers are given for some typical key questions.

Students will need to remember two points when using this book. Firstly, it covers all of the major 'O' level syllabuses, but these differ slightly in content. Most students will recognize those items which are not relevant to their particular syllabus and ignore them. If there is any doubt about syllabus content a chemistry teacher/lecturer should be asked for advice, or alternatively (or in addition) a copy of the syllabus produced by the appropriate Examining Board may be bought through a local bookseller or direct from the offices of the Board, whose addresses are given on page 8. A syllabus is also useful as a checklist during the revision programme.

Secondly, this Study Aid does not describe in detail the experimental procedures which lead to an understanding of the facts and principles involved in an 'O' level course. It is assumed that students will have followed a scheme involving some laboratory work, and that they have access to a notebook describing this work. If a topic in the syllabus could be examined by questions requiring a description of an experiment this is clearly indicated. By far the most difficult requirement of modern chemistry courses is the ability to understand and correlate principles, and to utilize these ideas in different situations. This Study Aid is concerned with these requirements, for although the experimental techniques and procedures are relatively easy to understand and to learn, the conclusions to be drawn from them present greater difficulty.

In order to make it easier for the student to recognize how practical work links with the facts and principles being discussed, typical experiments which may have been used are briefly referred to. It cannot be over-emphasized, however, that the experiments referred to are typical only, and that individual schools and colleges will have their own way of developing practical work. It is up to the student to link his or her own practical work with the topics in this book; this exercise alone will guarantee a greater understanding of the essential principles, for although individual experiments may differ, the conclusions will be the same.

The examination boards

The addresses given below are those from which copies of syllabuses and past examination papers may be ordered. The abbreviations (AEB, etc) are those used in this book to identify actual questions.

Associated Examining Board, (AEB)
Wellington House,
Aldershot, Hants. GU11 1BQ

University of Cambridge Local Examinations Syndicate, (CAM)
Syndicate Buildings, 17 Harvey Road,
Cambridge CB1 2EU

Joint Matriculation Board, (JMB)
(Agent) John Sherratt and Son Ltd.,
78 Park Road,
Altrincham, Cheshire WA14 5QQ

University of London School Examinations Department, (LOND)
66-72 Gower Street,
London WC1E 6EE

Northern Ireland Schools Examinations Council (NI)
Examinations Office,
Beechill House,
Beechill Road,
Belfast BT8 4RS

Oxford Delegacy of Local Examinations, (OX)
Ewert Place,
Summertown,
Oxford OX2 7BZ

Oxford and Cambridge Schools Examination Board, (O & C)
10 Trumpington Street,
Cambridge CB2 1QB

Scottish Certificate of Education Examining Board, (SCOT)
(Agent) Robert Gibson and Sons, Ltd.,
17 Fitzroy Place,
Glasgow G3 7SF

Southern Universities Joint Board, (SUJB)
Cotham Road, Bristol BS6 6DD

Welsh Joint Education Committee, (WEL)
245 Western Avenue,
Cardiff CF5 2YX

1 Atomic structure, kinetic theory and bonding

1.1 Atoms, molecules and ions

Mistakes are frequently made in using these terms, for students often write 'atom' when they should really be using either molecule or ion. Free atoms are rarely encountered in chemistry, the noble gases being the only ones to exist naturally as simple atoms. Each of these terms has a specific meaning and it is important to avoid using them incorrectly; if in doubt write 'particles'.

Ions are formed when atoms gain or lose electrons (negative ions and positive ions respectively) and molecules are formed when two or more neutral atoms join together by covalent bonding. It will be easier to use these terms correctly when bonding and structure are understood.

*An **atom** is the smallest particle into which an element can be divided without losing its identity.*
*A **molecule** is a group of atoms (held together by covalent bonding) which is capable of independent existence.*
*An **ion** is a charged particle formed from an atom or group of atoms by the loss or gain of electrons.*

Evidence for the existence of particles

It is not possible to *prove* the existence of these particles at school level, but many experiments provide indirect evidence. For example, coloured solutions can be diluted many times until only a very small proportion of the original substance is present, but the colour is still visible and is spread evenly throughout the container. We can explain this uniform disintegration of a substance if we believe that the original material consists of many separate particles, some of which are present (but in ever-decreasing concentrations) in each of the diluted solutions.

Diffusion experiments (Unit 1.3) are impossible to explain without accepting the existence of small particles. The fact that some substances (e.g. water) can exist in three different states (solid, liquid and gas), each with its own properties, can be understood if we visualize the presence of particles, packing differently and moving differently in each of the states of matter. You could be asked to describe an experiment which has helped to convince you that substances are made up of particles.

How small are atoms and molecules?

It is possible, even in a school laboratory, to determine the approximate size of a cluster of atoms, i.e. a molecule. A typical substance chosen for the investigation is stearic acid, which has molecules consisting of 56 atoms. (Its formula is $C_{17}H_{35}COOH$, and its sodium salt is common soap.)

A small measured volume of a very dilute solution of stearic acid in ethoxyethane (diethyl ether) is dropped onto a water surface previously sprinkled with a fine powder such as sulphur. The ethoxyethane quickly evaporates leaving the molecules of stearic acid (which float on top of the water) to 'tumble over each other' and spread out along the top of the water until they form a layer one molecule thick. The area of this monomolecular layer can be measured because it is visible as an 'oil slick' which has pushed the powder back to the edges of the slick. (Experimental details needed.)

As   volume (V) = area (A) × height or thickness (t)

then $$t = \frac{V}{A}$$

If this is applied to the oil slick, V and A are known and t can be calculated. In this case t is the thickness of the stearic acid layer, i.e. the diameter of a molecule of the acid. (*Note*: V is the volume of the stearic acid and not the volume of the acid plus ethoxyethane mixture used initially.) Experiments of this type give values of the order of 10^{-7} cm (i.e. the length obtained by dividing a centimetre into 10 equal parts, and then dividing one of these parts into 10 and so on until 7 divisions have been made altogether) for the diameter of a molecule. Chemists usually measure such small sizes in nanometres (nm); one nanometre is 10^{-9} metre. Atoms are smaller than molecules such as stearic acid, and although they vary in size, a typical atom would have a diameter of the order of 0.2 nm (0.2×10^{-9} metre or 2×10^{-8} cm).

1.2 Atomic structure

Atoms consist of a very small, dense nucleus in which are found protons (positively charged) and neutrons (neutral). Outside the nucleus are the electrons (negatively charged) which are found in energy levels. Many energy levels exist within an atom (although they are not always occupied), and each level can take up to a certain maximum number of electrons. At 'O' level we are normally concerned only with the structures of the first 20 elements, and for these purposes we can assume a maximum of 2 electrons in energy level 1, 8 in energy level 2 and 8 in energy level 3.

Particle	Approximate mass (atomic mass units)	Charge
proton	1.0	+1
neutron	1.0	0
electron	1/1840	−1

In order to draw the structure of a given atom we need to know two pieces of information, the atomic number and the mass number. These data are usually provided in examination questions by giving the symbol for the element together with two numbers, e.g. $^4_2$He. The usual convention is for the top number to be the mass number and the bottom one the atomic number. (If in doubt, where the two numbers differ the larger of them is always the mass number.) Thus $^4_2$He tells us that the mass number of helium is 4 and the atomic number is 2. Sometimes the only data needed are the mass number (or the relative atomic mass, which is explained on p. 13) and this is then given in the form $A_r(H) = 1$, $A_r(Na) = 23$, etc.; or simply H = 1, Na = 23, etc.

The atomic number of an atom is defined as the number of protons present in that atom.
The mass number of an atom is defined as the sum of the numbers of protons and neutrons in that atom.

The information obtainable from these two terms is summarized below.

Atomic number = number of protons
      (also = number of electrons in neutral atom)
Mass number = number of protons + number of neutrons
(Mass number − atomic number) = number of neutrons

An atom of $^{35}_{17}Cl$ thus consists of 17 electrons and 17 protons (because the atomic number is 17) and also 18 neutrons (because mass number − atomic number = 18), arranged as

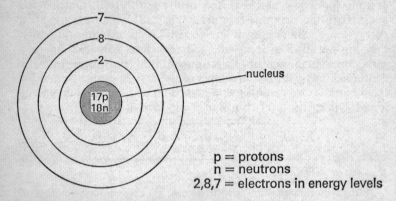

p = protons
n = neutrons
2,8,7 = electrons in energy levels

Note: When drawing atomic structures *always* label your diagram or use a key.

Isotopes

Most elements have more than one kind of atom, for it is possible to vary the number of neutrons in the nucleus while still retaining the fixed numbers of electrons and protons which are characteristic of the element. The variation in neutrons affects only the mass of the atom and has no influence on the chemical properties, which depend on the numbers of electrons and protons. All of the isotopes of one element therefore have identical chemical properties, because they have the same numbers of electrons and protons (i.e. the same atomic number).

Isotopes are atoms of the same element which contain different numbers of neutrons, i.e. they have the same atomic number but different mass numbers.

Isotopes of chlorine include $^{35}_{17}Cl$ and $^{37}_{17}Cl$:
As chlorine always contains these two isotopes in the ratio of 3 parts $^{35}_{17}Cl$ to 1 part $^{37}_{17}Cl$, the relative atomic mass of an 'average atom' of chlorine is $\frac{(3 \times 35) + 37}{4} = 35.5$, which is the relative atomic mass of chlorine and *not* the mass number.

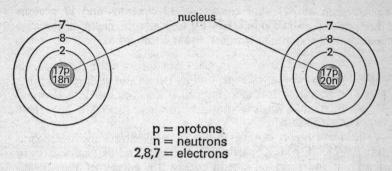

p = protons
n = neutrons
2,8,7 = electrons

Isotopes of chlorine

The relative atomic mass of an element is the mass of an 'average' atom of the element compared with that of an atom of $^{12}_{6}C$, the mass of which is taken as 12 units.

A mass number refers only to one particular isotope or type of atom, and is always a whole number. A relative atomic mass refers to an 'average atom' of the element (which may not actually exist) and may not be a whole number.

The relative molecular mass of a substance (M_r) is defined as the mass of an 'average molecule' of the element or compound relative to that of the mass of an atom of $^{12}_{6}C$, which is taken as 12 units.

A relative molecular mass is conveniently obtained by adding together the individual atomic masses of the atoms present in a molecule. For example, if the atomic masses are $A_r(H) = 1$, $A_r(S) = 32$ and $A_r(O) = 16$, then the relative molecular mass of sulphuric acid, H_2SO_4, is $(2 \times 1) + 32 + (4 \times 16) = 98$. This means that a molecule of sulphuric acid would 'weigh' 98 units on a scale where an atom of the $^{12}_{6}C$ isotope 'weighs' 12 units.

If a compound is ionic it will not contain molecules and we should refer to its formula mass rather than relative molecular mass. This is derived in exactly the same way, e.g. the formula mass of sodium chloride (NaCl) is $23 + 35.5$ (from Na = 23 and Cl = 35.5) = 58.5. It would be acceptable to say that this was the relative molecular mass of sodium chloride, but it is less confusing to call it the relative formula mass.

1.3 The kinetic theory

Kinetic energy

It is not intended that you should learn off by heart the work on kinetic theory, change of state, boiling points and gas pressure; this would be pointless. It is more important that you should *understand* the ideas, so that you can use them in later work. The important factors are indicated.

The kinetic theory assumes that at all temperatures above 'absolute zero' (−273 °C or 0 K) particles have energy and that this energy is largely in the form of kinetic energy, i.e. energy of movement. Kinetic energy can exist in several forms such as rotation, vibration and translocation.

Rotation (of the whole molecule)

Vibration (within a molecule)

Translocation (movement of a molecule from place to place)

Forms of kinetic energy

The energy is proportional to the temperature on the Kelvin scale (to change °C to K add 273), so that as a substance is heated its particles gain more and more kinetic energy. (Remember that a change of temperature from 20 °C to 40 °C is not a doubling of the temperature, as this is really only a change from 293 K to 313 K.)

Change of state

Most chemicals can exist in three different states of matter: solid, liquid and gas. These are interconnected by the following physical changes.

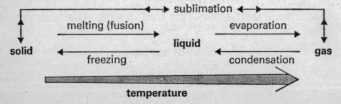

Atomic structure, kinetic theory and bonding

In a solid, the particles are packed tightly together and their kinetic energy is almost entirely in the form of vibrational and rotational movements. As the temperature increases the particles gain more and more kinetic energy and their movement increases until, at the melting point, the particles have enough energy to 'come apart' and to have translocational energy.

As the temperature increases further this tendency to move about increases (although the particles are still close together) until some molecules have so much energy that they escape from the liquid and have a much greater degree of freedom as a gas. Particles escape as vapour before the boiling point is reached, but at the boiling point a special set of conditions is reached and there is a very rapid change from liquid to gas. (At the boiling point the vapour pressure of the gas being formed in the liquid is equal to the atmospheric pressure above the liquid.) The boiling point of water is 100 °C or 373 K at a pressure of 760 mm mercury, but this changes if the pressure changes.

solid

Ordered structure, closely packed, kinetic energy restricted to vibrations and rotations. Intermolecular forces strong compared with kinetic forces

liquid

Less ordered structure, still closely packed but kinetic energy now in vibrations, rotations, and translocations. Intermolecular forces and kinetic forces of same order of magnitude

gas

No structure, molecules virtually independent and far apart, all movements rapid. Intermolecular forces now weak compared with kinetic forces

The states of matter

The molecules of a gas are moving very quickly in all three forms of movement, and are relatively far apart. This is why there is an enormous change in volume (about 1700 times) when a liquid changes into a gas.

Once a pure liquid has reached its boiling point the temperature will not rise any further (unless the pressure is changed) because this temperature represents the point at which molecules cannot receive any more energy (e.g. heat) without changing into a gas. The heat still being applied at the boiling point is being used to separate the molecules (break intermolecular forces) and not to raise their temperature; this is the latent heat of vaporization.

It should follow that gases can be condensed into liquids by cooling (which reduces the kinetic energy and thus increases the tendency of intermolecular forces to bring the molecules closer together) and by increasing the pressure (which compresses the molecules closer together). One or both of these factors will eventually produce the liquid state, and ultimately a solid.

Boiling points

If the pressure above a liquid which is being heated is higher than normal, then the liquid molecules have to 'fight' harder to escape from the liquid as a gas. When the pressure is high the molecules therefore need more kinetic energy before they can escape, and they need to be heated to a higher temperature before they boil. This is why the boiling point of a liquid depends upon the atmospheric pressure, and why the boiling point increases as the pressure increases. This explains how a pressure cooker cooks food more quickly than an open pan because the water inside is under pressure and can reach a temperature higher than 100 °C before boiling, whereas food being boiled in a pan open to the atmosphere cannot reach a temperature much above 100 °C. Similarly sulphur (melting point 119 °C) can be melted by 'superheated water' in the Frasch process (p. 150), and the boiling point of a liquid is lower than normal up a high mountain where the atmospheric pressure is lower. You could be asked to describe an experiment to show how the boiling point of water depends upon the atmospheric pressure.

Similarly, if a substance is dissolved in a liquid (i.e. the liquid is made impure) the surface area from which the liquid molecules can escape is reduced by the presence of 'foreign' particles.

As with increased atmospheric pressure, the molecules of liquid have less chance to escape (from a solution) and they must be supplied with more energy than when in the pure state before they can

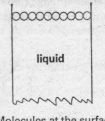
Molecules at the surface in a pure liquid

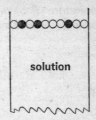
Molecules at the surface when the liquid contains an impurity

How the evaporation of liquids is influenced by dissolved substances

escape as easily. The boiling point of a pure liquid is thus lower than that of one of its solutions, and you could be asked to describe an experiment which shows this.

The boiling point of a liquid is raised if the atmospheric pressure is increased or if another substance is dissolved in it.

Gas pressure
The kinetic theory explains how gases exert pressure. The molecules, being in constant movement, bombard the walls of their container and this is responsible for the pressure. If the temperature of a certain mass of gas is increased, the molecules gain more kinetic energy, move more rapidly, and bombard the walls of the container more frequently and more vigorously. If the pressure is kept constant as the temperature rises, then the volume is increased; and if the volume is kept constant then the pressure is increased.

Charles' law states that the volume of a fixed mass of gas is directly proportional to its temperature (on the Kelvin scale) if the pressure remains constant.

Thus $P \propto T$ (in K, not °C), and if the Kelvin temperature is doubled then the volume of the gas is doubled.

If the volume of a certain mass of gas is reduced (at constant temperature), the molecules become more concentrated and the number of hits they make on a certain area of the container is increased, thus again increasing the pressure. If the volume is halved, then the pressure is doubled.

Boyle's law states that the volume of a fixed mass of gas is inversely proportional to the pressure, if the temperature remains constant, i.e. $P \propto \dfrac{1}{V}$.

These two laws are often combined in the general gas equation in the form

$$\frac{P_1 V_1}{T_1} = \frac{P_2 V_2}{T_2}$$

where P_1, T_1 and V_1 are the initial pressure, temperature (in Kelvin) and volume, and P_2, T_2 and V_2 the final ones.

Evidence for the movement of particles

(a) Diffusion

If two miscible liquids are carefully placed in a container so that there is at first a sharp boundary between them, they gradually mix together over a period of time until the two are evenly mixed, i.e. the solution is homogeneous. Each liquid is said to diffuse into the other.

Diffusion is the movement of a substance from a region of high concentration to one of a lower concentration.

As this mixing takes place without any heating or shaking, it suggests that the particles within the liquids are moving, i.e. have kinetic energy. As diffusion in gases is very rapid compared to that in liquids, this supports the theory that gas molecules have more kinetic energy and freedom of movement than those in liquids. You could be asked to describe experiments on diffusion in the gas and liquid states.

(b) Brownian movement

If small pieces of solid, each too small to be seen with the naked eye but big enough to be seen under a microscope, are suspended in a liquid or gas, they can be seen to be darting about rapidly in a random manner. At school level examples of this type might include pollen floating on water, small pieces of carbon suspended in the air (smoke), or pieces of carbon suspended in a liquid.

This movement is caused by the rapid bombardment by the (invisible) air or water molecules which surround the small pieces of solid. It is most unlikely that at any given moment an exactly equal number of molecules of water (or air) will hit each piece of solid on all sides; greater bombardment from one side pushes the piece of

Atomic structure, kinetic theory and bonding 19

solid along, but a split second later it is sent in another direction. You could be asked to describe an experiment which shows this effect.

***Brownian movement** is the random movement of microscopic pieces of solid caused by irregular bombardment from the molecules which surround them in the gas or liquid states.*

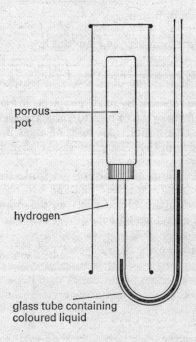

To illustrate diffusion

(c) Gas molecules diffuse at different speeds
If a beaker of hydrogen is placed over a porous pot containing air and connected to a manometer (see Figure), the manometer shows that hydrogen diffuses into the pot faster than the air diffuses out. The lighter hydrogen molecules diffuse more quickly than the heavier air molecules. (Details may be required of an experiment which shows that one gas diffuses faster than another.) Similarly, molecules of carbon dioxide (more dense than air molecules) diffuse into the pot more slowly than air diffuses out.

Graham's law of diffusion *states that the rate of diffusion of a gas (at constant temperature and pressure) is inversely proportional to the square root of its density, i.e.* rate of diffusion $\propto \sqrt{\dfrac{1}{density}}$.

(Only a few syllabuses include this law.) As density (in gases) is proportional to molecular mass, we can also state:

$$\text{rate of diffusion} \propto \sqrt{\dfrac{1}{\text{molecular mass}}}.$$

Another useful expression derived from this law is

$$\dfrac{\text{rate of diffusion of A}}{\text{rate of diffusion of B}} = \sqrt{\dfrac{\text{density of B}}{\text{density of A}}}.$$

As densities can be replaced by relative molecular masses simply by multiplying the density by 2 (see p. 216), this expression can be used to determine the relative molecular mass of a gas if its rate of diffusion is compared with that of a gas whose relative molecular mass is known.

Another typical experiment involves a glass tube (see Figure) into which ammonia gas diffuses at one end and hydrogen chloride gas (from concentrated hydrochloric acid) at the other. If you are familiar with this, experimental details may be required. Where the

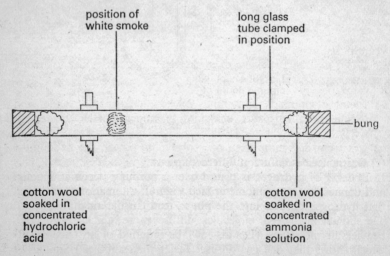

To compare the rate of movement of two gases

two gases meet, a white smoke of ammonium chloride is formed ($NH_3(g) + HCl(g) \rightarrow NH_4Cl(s)$) and this is found to occur not in the middle, but nearer to the end where the hydrogen chloride started. The lighter molecules of ammonia (molecular mass 17) have diffused further than the heavier hydrogen chloride molecules (relative molecular mass 36.5) in the same time.

The time taken for the gases to meet is longer than one would expect from a knowledge of their velocities, which are very high indeed even at room temperature. This is because the molecules travel in a random way (not in a straight line) due to constant collisions with other molecules of the same kind and with air molecules also present. This factor influences the rate of diffusion of any gas in air, and explains why diffusion is very rapid in a vacuum.

1.4 Symbols, formulae and bonding. Molecular shapes

You need to know the symbols and combining powers (**valencies**) of all the common elements and radicals (given in Table 1.1). A radical is a group of atoms which usually forms the non-metallic part of a compound and which can remain intact during a chemical reaction, behaving in many ways like a single atom and always having the same combining power. The sulphate group, SO_4, is a radical. Radicals cannot exist on their own. Make sure that you know how to work out a chemical formula from the appropriate symbols and combining powers.

When writing the **name of a chemical**, remember the following points.

1 Metal compounds containing oxygen, chlorine, bromine, and iodine are called oxides, chlorides, bromides, and iodides respectively.
2 If a metal has more than one combining power, the name must clearly indicate which power is being used in the compound. This is done by placing the appropriate combining power in Roman numerals inside a bracket immediately after the element to which it belongs, e.g. copper(II) sulphate, iron(III) chloride.

(It is *not* necessary to show the combining power of such an element when writing a *formula*, because the formula itself enables us to work out the combining power being used, e.g. $FeCl_3$ must mean iron(III) chloride and cannot mean iron(II) chloride.)

Table 1.1 *Symbols and combining powers of the common elements and radicals*

Note: The elements in italic have more than one combining power.

	Symbol	Stable ion formed
Elements and radicals with a combining power of 1		
chlorine (chloride)	Cl	Cl^-
bromine (bromide)	Br	Br^-
iodine (iodide)	I	I^-
hydroxide	OH	OH^-
hydrogencarbonate	HCO_3	HCO_3^-
hydrogensulphate	HSO_4	HSO_4^-
nitrate	NO_3	NO_3^-
manganate(VII) (permanganate)	MnO_4	MnO_4^-
copper(I)	Cu	Cu^+
silver	Ag	Ag^+
sodium	Na	Na^+
potassium	K	K^+
ammonium	NH_4	NH_4^+
hydrogen (hydride)	H	H^+ or H^-
Elements and radicals with a combining power of 2		
oxygen (oxide)	O	O^{2-}
sulphate	SO_4	SO_4^{2-}
sulphite	SO_3	SO_3^{2-}
carbonate	CO_3	CO_3^{2-}
sulphur (sulphide)	S	S^{2-}
dichromate(VI)	Cr_2O_7	$Cr_2O_7^{2-}$
lead(II)	Pb	Pb^{2+}
zinc	Zn	Zn^{2+}
tin(II)	Sn	Sn^{2+}
magnesium	Mg	Mg^{2+}
calcium	Ca	Ca^{2+}
copper(II)	Cu	Cu^{2+}
iron(II)	Fe	Fe^{2+}
Elements and radicals with a combining power of 3		
phosphate	PO_4	PO_4^{3-}
aluminium	Al	Al^{3+}
iron(III)	Fe	Fe^{3+}
Elements and radicals with a combining power of 4		
carbon	C	no ion
silicon	Si	no ion
lead(IV)	Pb	Pb^{4+}
tin(IV)	Sn	Sn^{4+}

Bonding

Only atoms of the noble gases exist in an uncombined state. Other atoms *usually* react in such a way as to gain stable electron structures (usually that of the nearest noble gas, i.e. a fully filled outer shell of electrons) by losing, gaining, or sharing electrons. They do this by joining (bonding) with other atoms, even if this means atoms of their own kind.

There are two main types of bonding, **ionic** (or electrovalent) and **covalent**. When atoms form ions in order to combine, the bonding is ionic. When atoms share electrons in joining together, the bonding is covalent.

Ionic bonding

If a metal or the ammonium radical is combined with a non-metal or radical, the bonding is likely to be ionic.

This arises because metals usually have 1, 2, or 3 electrons in their outer energy levels and thus find it comparatively easy to *lose* electrons in order to attain the electron structure of a noble gas. Non-metals, on the other hand, need to *gain* electrons to attain stable electron structures, and if a metal is there to donate them the non-metal will receive them. The metal produces a positive ion by loss of one or more electrons, and the non-metal a negative ion by gaining one or more electrons. Not only are both elements 'satisfied' by this loss and gain of electrons, they are then held together by the opposite electrical charges of the ions produced. The rules which follow should enable anyone to explain how atoms combine by ionic bonding. If you are already confident about your ability to do this, move on to the next section.

In order to show how atoms combine by ionic bonding, proceed as follows. (An example using magnesium chloride is given after the procedure, on page 25, and this should be consulted at each stage.)

1 Determine the formula of the substance, e.g. $MgCl_2$.
2 Work out the atomic structures (excluding neutrons) of the atoms concerned from their atomic numbers.
3 Draw the atomic structure of each of the atoms in the formula (e.g. for $MgCl_2$ draw one atom of magnesium and two atoms of chlorine), keeping the metal atoms on the left hand side. (If more than one atom of the same kind is needed, draw the additional atoms in line below the first one of the same type.) Draw the *outer*

electrons in the atoms individually, using different colours or different symbols for those in the metal atom(s) and those in the non-metal atom(s).
4 Show by means of one or more arrows how the outer electrons in the metal atoms move over to the outer energy levels of the non-metal atoms, so as to leave the original outer shells of the metal atoms empty and to fill the outer shells of the non-metals. (If you have worked out the formula correctly, then the transfer of the appropriate number of electrons should be automatic.)
5 Label fully, as shown in the example.
6 Redraw the 'atoms' as they will be after the electron transfer, showing the individual electrons in the outer shells of the non-metal ions with their different colours or symbols. Only one example of each type of ion need by drawn, but the numbers of each should be clearly shown.
7 Put a bracket and charge symbol around each ion. Remember that if an atom loses electrons it becomes a positively charged ion, with the same number of positive charges as the number of electrons lost. This is because the protons now outnumber the electrons. The opposite is true of non-metals, which gain electrons. Metals always form positive ions.
8 Write the name and formula of the compound formed, and label as shown; it is not necessary to label the nucleus or electrons in the second drawing as they were labelled in the first.

The example on the following page, using magnesium chloride, illustrates these steps.

Other examples to try include sodium chloride, potassium chloride, potassium fluoride, sodium oxide, calcium fluoride, magnesium oxide and aluminium fluoride. These examples cover all types of ionic bonding.

Properties of ionic compounds
1 Ionic compounds are composed of two or more different kinds of oppositely charged ions.
2 These oppositely charged ions attract each other and form a large three-dimensional lattice, called a giant structure, which is held together by electrostatic attractions in all directions. Ionic compounds are thus usually crystalline solids.
3 Because of the great attraction between the ions, a large amount of energy has to be used to separate them, and ionic compounds usually have high melting points and boiling points.

Atomic structure, kinetic theory and bonding 25

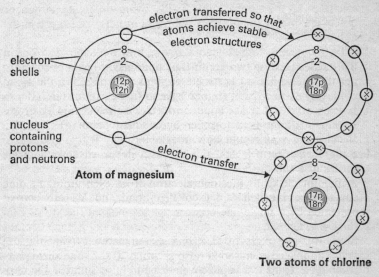

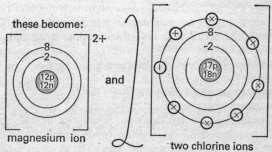

Magnesium chloride, MgCl$_2$

The formation of magnesium chloride

4 Ionic compounds, when molten or in aqueous solution, are electrolytes.
5 Ionic compounds are usually soluble in water but do not dissolve in organic solvents such as ethanol or benzene.
6 Once ions are formed they show none of the chemical properties of the parent atoms.
7 Ionic bonds are non-directional (see properties of covalent compounds, p. 30).

Covalent bonding

If two or more non-metals combine together, the bonding is likely to be covalent.

This arises because two non-metals both need to gain electrons in order to achieve noble gas structures. This time there is no metal to provide any electrons and so they have to share electrons in order to help each other. They are bonded together by the shared electrons. A group of atoms held together by covalent bonding is called a molecule, and you should now understand why it is important to use the correct term (atom, ion or molecule) when describing the particles of a substance.

A pair of electrons (i.e. one electron from each atom) binding atoms together is called a single covalent bond, and this is sometimes represented by a single line between the symbols of the atoms concerned, e.g. Cl—Cl.

Similarly, two pairs of electrons shared between two atoms is called a double covalent bond (also shown as C=C) and three pairs of shared electrons is a triple covalent bond (also shown as N≡N). The rules which follow should enable anyone to explain how atoms form covalent bonds to produce a named molecule. If you are already confident about your ability to do this, move on to the next section.

In order to show how atoms combine by covalent bonding, proceed as follows. At first these steps will seem complicated, but you should follow each stage using a simple molecule such as the chlorine molecule, the completed diagram for which is shown on page 28. With practice you will be able to combine some of the steps.

1 Determine the formula of the molecule in question.
2 Draw atomic structures for each of the atoms in the molecule, setting them out as follows. (Inner electron shells can be shown by numbers on the circles, but at this stage draw only a circle for the outer shell of each atom.) If there are only two atoms in the molecule draw them side by side, with outer shells overlapping. If there are three or more atoms in the molecule, place the atom with the largest combining power (i.e. the one with the smallest number of atoms in the formula) in the centre of the page, and distribute the other atoms evenly around it so that their outer shells overlap the outer shell of the central atom. Thus for an ammonia molecule, NH_3, the atoms would be spaced like this:

Atomic structure, kinetic theory and bonding 27

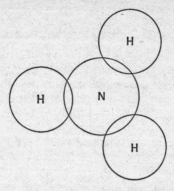

If the atom with the largest combining power occurs twice, draw the two atoms alongside each other (with outer shells overlapping) in the centre of the page and divide the other atoms evenly around them, not touching each other. Thus for ethane, C_2H_6, the atoms would be spaced like this:

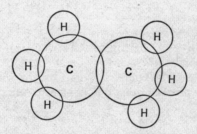

3 Using a pencil, lightly mark in the individual electrons in each outer energy level, using different symbols or colours for each atom and keeping them away from areas where atoms touch.
4 Draw a box around each of the areas where outer shells touch (see diagrams overleaf).
5 Move to each box (by erasing and redrawing) one electron from each of the atoms which form the box. Each box will now contain a pair of electrons shared by both atoms touching the box. (Retain your original symbols for the electrons.)
6 If necessary move more electrons (always working in pairs, one

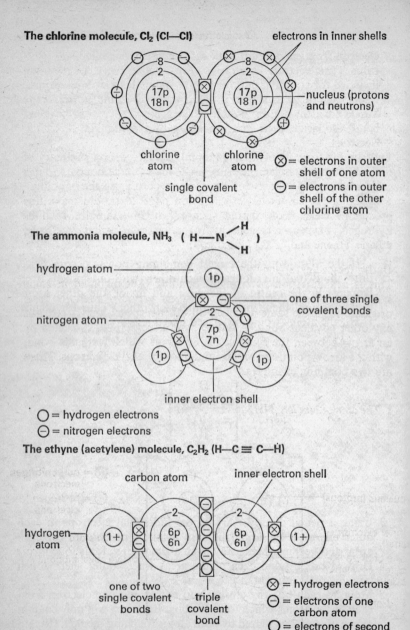

Some covalently bonded molecules

electron from each atom) until each outer shell is filled with electrons. (The electrons in each box are counted twice, i.e. they are included for counting purposes in the outer shells of both atoms.)

7 Label each atom as shown in the completed examples, and write 'single covalent bond', 'double covalent bond' or 'triple covalent bond' alongside each box according to the number of shared electrons.

8 Write the formula of the compound together with a summary of the bonding arrangement using single lines (single bonds), double lines or triple lines between the atoms according to the bonding.

Studying the completed examples on page 28 should make this procedure clear. You should then try other examples such as hydrogen, oxygen, nitrogen, hydrogen chloride, water, methane, ethane, ethene and carbon dioxide.

Dative (co-ordinate) bonding

This is a subdivision of covalent bonding (not included in some syllabuses) where both of the electrons forming a shared pair originate from the same atom. In all other respects, this is just like any other covalent bond. A dative bond is sometimes indicated by an arrow between the symbols of the atoms which form the bond, with the arrow pointing away from the source of the electrons. There are two common examples.

1 *The ammonium ion*, NH_4^+,
$$\left[\begin{array}{c} H \\ | \\ H-N-H \\ \downarrow \\ H \end{array} \right]^+$$

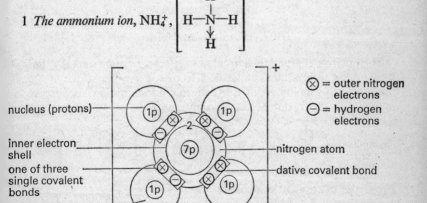

The ammonium ion, NH_4^+

The positive charge arises because in effect a positive ion (H^+) has been added to a neutral molecule (NH_3). A typical example of the formation of this ion is when ammonia gas and hydrogen chloride gas come into contact, i.e. (using outer electrons only):

$$H \overset{\cdot\times}{\underset{H}{:}} \overset{H}{\underset{\cdot\times}{N}} {\times} + H \overset{\circ\circ}{\underset{\circ\circ}{\circ}} Cl \overset{\circ\circ}{\underset{\circ\circ}{\circ}} \rightarrow \left[H \overset{\cdot\times}{\underset{H}{:}} \overset{H}{\underset{\cdot\times}{N}} {\times} H \right]^+ + \left[\overset{\circ\circ}{\underset{\circ\circ}{\circ}} Cl \overset{\circ\circ}{\underset{\circ\circ}{\circ}} \right]^-$$

2 *The hydronium ion*, H_3O^+, $\left[\begin{matrix} H \\ H \end{matrix} \!\!\!> O \rightarrow H \right]^+$

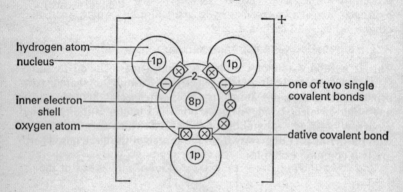

The hydronium ion, H_3O^+

The positive charge arises because in effect a positive ion (H^+) has been added to a neutral molecule (H_2O). A typical example of the formation of this ion is when hydrogen chloride molecules dissolve in water, i.e. (using outer electrons only):

$$H \overset{\times\cdot}{\underset{\times\times}{:}} \overset{H}{\underset{}{O}} {\times} + H \overset{\circ\circ}{\underset{\circ\circ}{\circ}} Cl \overset{\circ\circ}{\underset{\circ\circ}{\circ}} \rightarrow \left[H \overset{\times\cdot}{\underset{\times\times}{:}} \overset{H}{\underset{}{O}} {\times} H \right]^+ + \left[\overset{\circ\circ}{\underset{\circ\circ}{\circ}} Cl \overset{\circ\circ}{\underset{\circ\circ}{\circ}} \right]^-$$

Properties of covalent compounds
1 Covalent compounds consist of two or more atoms linked together by covalent bonds to form individual molecules.
2 Although the covalent bonds *inside* each molecule (the intramolecular bonds) are very strong and cannot easily be broken, the forces acting *between* the molecules (the intermolecular forces)

Atomic structure, kinetic theroy and bonding 31

are weak and relatively easy to break, e.g. dipole-dipole forces, or van der Waals forces.

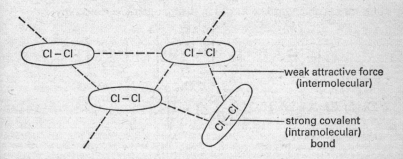

Intermolecular and intramolecular forces

This means that molecules are easily separated, i.e. *molecular* substances have low boiling points and melting points. In the example above chlorine is a gas at room temperature because it has 'already' melted and boiled at this temperature, and the weak intermolecular forces have been made insignificant compared with the kinetic energy of the molecules; the molecules have separated and enjoy the independence and freedom of movement of the gas state.

Note that silicon oxide, diamond and graphite are covalently bonded but they exist as giant structures rather than molecules. The points about boiling points, melting points, intermolecular forces and intramolecular bonds apply only to those covalent substances which are molecular, i.e. to most covalent materials but not all.

3 Covalent compounds are often insoluble in water, but dissolve more readily in organic solvents.
4 Covalent compounds do not conduct electricity when molten or dissolved in water.
5 Covalent bonds are directional (i.e. they 'lock' atoms in certain positions and at certain angles) unlike ionic bonds, where the positive or negative charges are distributed evenly around the whole ion. This means that covalent molecules have a certain shape, and this is discussed under the next heading.

Shapes of covalent molecules

The methane molecule, CH_4 This is tetrahedral in shape:

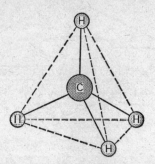

The shape of a methane molecule

The H—C—H bonds are all equivalent and are approximately 109°.

The ammonia molecule, NH_3 This has the shape of a triangular-based pyramid:

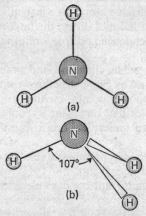

The arrangement of the atoms in a molecule of ammonia. The arrangement (a) does not show the shape of the molecule. The pyramidal shape of the molecule is illustrated in (b).

The H—N—H angles are approximately 107°, slightly less than the tetrahedral angle.

Atomic structure, kinetic theory and bonding 33

The water molecule, H_2O This is non-linear:

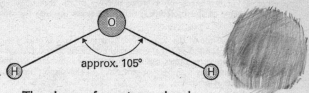

The shape of a water molecule

The H—O—H angle is approximately 105°.

The carbon dioxide molecule, CO_2, and the ethyne molecule, C_2H_2
These molecules are both linear:

The shape of a carbon dioxide molecule

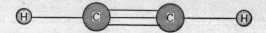

The shape of a molecule of ethyne

The ethene molecule, C_2H_4 This is a flat molecule (all atoms in the same plane), with each H—C—H angle 120°:

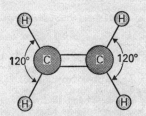

The shape of a molecule
of ethene (ethylene)

Questions

1 When pollen grains are suspended in water and viewed through a microscope, they appear to be in a state of constant but erratic motion. This is due to: (*a*) convection currents; (*b*) small changes in pressure; (*c*) small changes in temperature; (*d*) the bombardment of the pollen grains by molecules of water; (*e*) a chemical reaction between the pollen grains and the water.

34 Chemistry

2 An increase in temperature causes an increase in the pressure of a gas because: (a) it increases the mass of the molecules; (b) it causes the molecules to combine together; (c) it decreases the average velocity of the molecules; (d) it increases the average velocity of the molecules; (e) it increases the number of collisions between the molecules.

3 Isotopes are atoms of the same element which have: (a) the same mass number but different atomic numbers; (b) the same mass number and the same atomic number; (c) different atomic numbers and different mass numbers; (d) the same number of neutrons but different numbers of protons; (e) the same number of protons but different numbers of neutrons.

4 Sodium atoms and sodium ions: (a) are chemically identical; (b) have the same number of electrons; (c) have the same number of protons; (d) both react vigorously with water; (e) react and form compounds with covalent bonds.

5 An ionic bond is often formed when: (a) the combining atoms need to lose electrons in order to attain a noble gas configuration; (b) the combining atoms both need to gain electrons in order to attain a noble gas configuration; (c) two non-metallic elements react together; (d) a metallic element combines with a non-metallic element; (e) two metallic elements react together.

6 The elements W, X, Y and Z have atomic numbers, respectively, of seven, nine, ten and eleven. Write the formula for the compound you would expect to form between the following pairs of elements and indicate the type of bonding present: (a) W and X; (b) X and Z; (c) X and X; (d) Y and Y; (e) Z and Z.

7 Two plugs of cotton wool, one soaked in concentrated hydrochloric acid and the other in a concentrated solution of ammonia, are used to seal the ends of a horizontal glass tube. After a time a white ring forms nearer to one end of the tube than the other. Explain the formation and position of the ring in terms of the movement of gaseous molecules. (JMB)

8 In which substances would you expect to find free atoms at room temperature?

9 Chlorine has an atomic number of seventeen and exists mainly in isotopic forms, A and B, of atomic masses thirty-five and thirty-seven respectively. State the number of: (a) electrons in each atom of A; (b) protons in each atom of B; (c) neutrons in each atom of B; (d) electrons in each $^{35}_{17}Cl^-$ ion.

Calculate the A:B ratio in ordinary chlorine gas (molecular mass = 71). (JMB)

10 Name the three types of particles occurring in most atoms and give their relative masses and charges.

What do you understand by (a) the mass number, and (b) the atomic number of an element?

Draw a diagram showing the electronic structure of an atom, the nucleus of which carries a positive charge of nine units. What will be the valency of this element?

Explain what is meant by a covalent bond. Illustrate your answer by an example of a compound containing one or more covalent bonds. (AEB)

11 (a) The atomic number and atomic mass of nitrogen are seven and fourteen, respectively. Give a labelled diagram showing the structure of this atom.

(b) Indicate the relative charges and masses of the particles making up the nitrogen atom.

(c) Ammonia is a covalent compound. Give a diagram showing how the atoms in the ammonia molecule are bonded.

(d) Outline how you would prepare and collect a specimen of dry ammonia. (AEB)

12 Fluorine (9); Neon (10); Sodium (11); Magnesium (12).
    F   Ne   Na   Mg

The following refer to the above four elements, the atomic numbers of which are shown in brackets:

Atomic structure, kinetic theory and bonding 35

(a) What *two* facts about the structure of the atom of any *one* of these elements can be deduced from its atomic number?

(b) State the numbers of electrons in successive electron shells of the magnesium atom.

(c) If the symbol for the sodium ion is written Na^+, write similar symbols for the ions of fluorine and magnesium.

(d) Write the chemical formula for (i) magnesium fluoride; (ii) sodium fluoride.

(e) Name and explain briefly, with the aid of a diagram, the type of chemical bond linking atoms of fluorine in the molecule F_2.

(f) Caesium and sodium are both alkali metals. In what respect do the structures of atoms of these two elements resemble each other? (WEL)

2 Elements, compounds and mixtures. Purification

2.1 Elements, compounds and mixtures

Elements
*An **element** is a substance which cannot be simplified by any chemical process, i.e. all of the atoms within it have the same atomic number.*

Over a hundred elements are now known. These are the 'building bricks' of all substances. The elements can be divided into two main groups – metallic elements and non-metallic elements – although some elements have the properties of both. The main distinctions between metals and non-metals are summarized in Table 2.1. Although *most* metals will show *most* of the metallic properties listed, there are always exceptions and no single property provides any conclusive evidence as to whether an element is a metal or a non-metal.

Table 2.1 *Differences between metals and non-metals*

Metals	Non-metals
1 Good conductors of electricity	Poor conductors of electricity, except carbon in the form of graphite
2 Good conductors of heat	Poor conductors of heat
3 When freshly cut, have a shiny surface	Are often dull, even when freshly cut
4 Can be beaten into shape (are malleable) and drawn into wire (are ductile)	Usually shatter or break when treated in this way, i.e. are brittle
5 Sometimes liberate hydrogen from dilute acids	Never liberate hydrogen from dilute acids
6 Always form positive ions	Never form positive ions except for H^+ and NH_4^+
7 Usually have 1, 2 or 3 electrons in their outer energy levels	Usually have 5, 6, 7 or 8 electrons in their outer energy levels
8 Their oxides are usually basic	Their oxides are never basic, and are usually acidic or neutral

Compounds and mixtures

A compound is a substance composed of two or more elements which are joined together and which cannot be separated by physical processes.

(Physical processes include melting, boiling, dissolving, using a magnet, and the purification techniques discussed in 2.2, and contrast with chemical changes where bonds are broken and new ones made.)

Most of the chemicals with which you are familiar contain more than one kind of atom joined together and therefore are compounds. Remember that bronze, brass and other alloys are mixtures, not elements.

A mixture consists of two or more different substances which are not chemically combined, and which can therefore be separated by physical processes (e.g. standard purification techniques).

A mixture can consist of two or more elements (e.g. brass contains zinc and copper), two or more compounds (e.g. copper sulphate solution contains water and copper sulphate) or a combination of one or more elements and compounds (e.g. a mixture of zinc and sodium chloride). The differences between compounds and mixtures are summarized in the Figure on page 38 and in Table 2.2.

Table 2.2 *Differences between compounds and mixtures*

Compound	Mixture
1 Always homogeneous, i.e. perfectly uniform	May be homogeneous (e.g. water and ethanol) or heterogeneous (e.g. iron and sulphur)
2 Always contains the same elements in a fixed and definite ratio, i.e. has a constant composition (see the Law of Constant Composition, p. 103)	Can have any composition, e.g. from 1% sulphur and 99% iron, to 99% sulphur and 1% iron
3 The melting and boiling points are fixed, although the boiling point may vary with atmospheric pressure	The melting and boiling point depend upon the composition of the mixture. A mixture of 1% X and 99% Y could be called 'impure Y' and will have a melting point below that of pure Y. A mixture of 3% X and 97% Y will have an even lower melting point
4 Cannot be split up by physical processes	Separation is possible by physical processes, such as purification techniques
5 Its properties are totally unlike those of the elements within it	The substances in the mixture retain their own properties, and the mixture has all of these properties

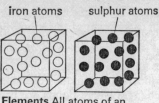

Elements All atoms of an individual element are alike

Mixture Atoms of the individual elements retain their own identities

Compound Individual 'units' composed of two atoms in this case

An illustration of the differences between elements, mixtures, and compounds

Point 5 in the table has important consequences in chemistry. When a compound is formed from its elements, the elements are *chemically combined* together and no longer retain their individual reactions. The compound formed has a completely new set of reactions. Consider the two elements sodium and chlorine. Sodium is a soft grey metal which reacts very vigorously with water. Chlorine is a poisonous gas which reacts to a small extent with water to produce an acidic solution which bleaches damp blue litmus paper. However, the compound formed from these two elements, sodium chloride (common salt), shows no metallic properties, does not react with water, and you 'eat' it every day in or with your food.

You could be asked to describe an experiment which shows the differences between elements, mixtures and compounds; a typical example would be one involving iron, sulphur and iron sulphide. If you have done an experiment of this kind, make sure that you can use it to make the points given in Table 2.2.

You will be expected to recognize whether any substance you have used in chemistry is an element, a compound or a mixture.

2.2 Purification techniques

When chemicals are made in the laboratory they are likely to be mixtures rather than pure substances, i.e. they will probably contain one or more impurities. Chemists regularly use physical processes to remove impurities or to separate one substance from another. You will be expected to understand the following techniques, and to recognize which of them would be useful in a particular situation.

Solution, filtration and crystallization

When describing processes of this type it is advisable to use as many of the key words (in bold type) as is conveniently possible.

A suitable **solvent** is chosen which (when hot if necessary) will **dissolve** some of the material in the mixture (the **solute**) to form a **solution** leaving the **insoluble** material in **suspension**. When the suspension is filtered the insoluble material is trapped on the filter paper as the **residue** and the clear **filtrate** is collected. The filtrate is heated (using a water bath or electrical heating if flammable) to **evaporate** some of the solvent, and when the concentrated solution is allowed to cool the solute **crystallizes** out, separated from the original mixture.

If the residue is required, it is washed and dried. If the solute is required, it also is washed (with a *little* solvent) and dried.

Typical applications are the purification of rock salt (using water as the solvent) and the preparation of many soluble salts (as in Unit 4.3). Solvents other than water may be needed for some mixtures. Centrifuging is an alternative to filtration, especially when working on a small scale.

Simple distillation
Distillation is the boiling of a solution in an apparatus so that the vapour is condensed and collected in a container separate from the original solution. The liquid collected in this way is called the distillate.

The essential principle is that the various components of a solution have different boiling points. When the mixture is heated, molecules of the more volatile component (i.e. the one with the lowest boiling point) will be the first to receive enough kinetic energy to escape from the liquid as a vapour. If this vapour is allowed to leave the distillation flask before being condensed back into liquid, it is separated from the original mixture.

You will have used simple apparatus in order to distil solutions such as copper sulphate solution and ink, and for more complex mixtures such as crude oil, but you should realize that if the distillation is to be efficient a Liebig condenser should be used.

Simple distillation is used to separate a solvent from a solution of a solid in a liquid (e.g. pure water from ink, copper sulphate solution, or sea water), or a liquid from a mixture of liquids. The efficiency of the process is limited, however, by the fact that it is quite possible for molecules other than the most volatile ones to evaporate from the solution. This becomes more important if the boiling points of the various substances are quite close. Even when a mixture of water and ethanol is distilled (boiling points 100 °C and 78 °C respectively) the distillate will contain both ethanol and water, although the ethanol is much more concentrated in the distillate than it is in the initial mixture. In such cases fractional distillation is more efficient.

Table 2.3 *Typical fractions obtained from the laboratory distillation of crude oil*

Fraction	Boiling range/°C	Colour	Viscosity	Flammability
petrol	30–65	colourless	very mobile (i.e. flows easily)	very flammable, clean flame
paraffin	65–175	colourless or pale yellow	mobile	flammable, yellow flame, some smoke
light oil	175–275	yellow	slightly viscous	more difficult to ignite, yellow flame, more smoke
lubricating oil	275+	dark yellow/brown	more viscous	difficult to ignite, very smoky flame, carbon residue

Note: The residue usually consists of heavy fuel oil and bitumen, which are difficult to separate in a school laboratory. The number of fractions obtained, and their properties, will depend upon the sample of oil used and experimental technique.

Elements, compounds and mixtures. Purification 41

The laboratory experiment to distil crude oil is often called fractional distillation, but strictly speaking no fractionating column is used and it is better regarded as a simple distillation in which successive fractions distil off as the temperature is raised. We are not concerned with an efficient separation in this case, for the fractions are themselves mixtures. Distillation of crude oil at a refinery does involve a fractionating 'column', for although the fractions obtained are still mixtures, the range of chemicals present in each mixture needs to be carefully controlled.

The main differences between the fractions obtained in a typical laboratory experiment to simplify crude oil by distillation are summarized in Table 2.3, and experimental details may be required.

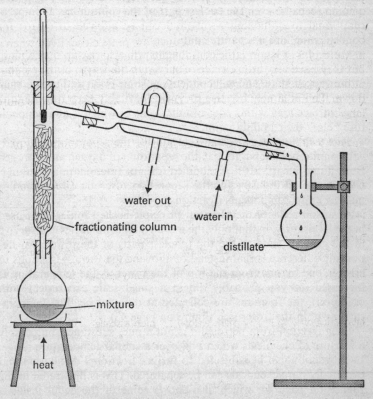

Fractional distillation of a mixture of two miscible liquids

Fractional distillation

This is a more efficient method of distillation, which involves the use of a fractionating column (see Figure, p. 41). The column is usually packed with glass beads or similar inert materials which provide a large surface area. When a mixture is distilled the column gradually warms up. There is a temperature gradient, i.e. the lower part of the column is warmer than the top. At first the vapour evaporating from the solution simply condenses on the inert surfaces in the column and trickles back into the flask. As the column gets hotter, the molecules in the vapour state rise higher before condensing, and eventually the top of the column reaches the temperature at which the most volatile component of the mixture boils. Molecules of this component can now 'survive' as gas molecules all the way up the column (because even the coolest part of the column has a temperature equal to its boiling point) and can be allowed to escape for condensation into a separate container.

Molecules of less volatile components will also rise up the column, but they have very little chance of surviving the length of the column in the vapour state, and will condense at some point and trickle back down. This is largely because they will inevitably keep touching the large surface area within the column, some of which is at a temperature below their boiling point.

This technique is particularly useful for the separation of liquids with similar boiling points, e.g. the separation of oxygen and nitrogen from liquid air (p. 210). A modified form of fractionating column is used in the oil industry for the separation of crude oil (a complex mixture) into components (fractions).

At an oil refinery the essential principle is the same as that used in the laboratory, in that only the more volatile molecules can survive in the gaseous state to the top (coolest part) of the tower. An important difference is that instead of allowing the various fractions to emerge one by one from the top of the tower as the temperature is increased (as you probably did in a small scale experiment with crude oil), the fractions are collected at different heights (temperatures) within the tower (see Figure on page 43).

Each fraction from petroleum is not a pure substance, but simply a mixture of chemicals which boil over a similar temperature range. Each fraction can be subjected to further fractional distillation and eventually pure chemicals can be separated. This is illustrated in the Figure on page 44, which also shows some of the detail inside a fractionating tower.

Elements, compounds and mixtures. Purification 43

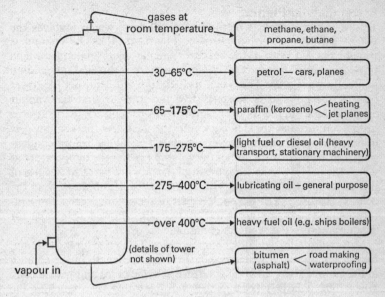

Fractional distillation of crude oil

The separation of immiscible liquids
Liquids which do not mix are described as immiscible, unlike those separated by distillation, which are miscible. You must never describe *one* liquid as being immiscible; the term applies only to a named *pair* of liquids. Thus water is miscible with ethanol but immiscible with oil. A separating funnel is used to separate immiscible liquids.

Chromatography
The mixture to be separated is placed at one edge of a stationary surface, over or through which a solvent is allowed to flow. Examples of the stationary surface include paper (cellulose) and aluminium oxide.

In elementary work the stationary surface is usually paper, and the solvent a liquid such as water or ethanol. The paper may take the form of a disc, in which case the mixture and the solvent are spotted on to the centre. Alternatively ascending chromatography may be used, in which spots of the mixture are made near the bottom of a paper strip which is then allowed to stand in a small volume of solvent in a bottle or gas jar. You should be able to describe experi-

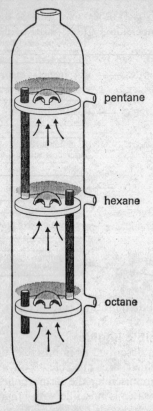

Secondary fractional distillation
of a fraction from a primary distillation

mental details of the use of chromatography to separate pigments such as those in ink, felt-tipped pens and green leaves, using appropriate solvents and either or both of the above techniques. You may also have used chromatography in determining which sugars are found in the natural polymer starch (p. 184) in which case the positions of the various spots can only be determined after spraying with a suitable locating agent.

Sublimation

It is sometimes possible to separate one solid from a mixture of solids by using the fact that one of them will sublime, i.e. change straight from solid to vapour when heated, and then reform the solid (separated from the mixture) in a cooler part of the apparatus (e.g. the top of a test-tube or inside a filter funnel held over a test-tube).

This method is greatly restricted by the fact that very few solids sublime, and elementary applications are largely confined to the separation of ammonium chloride or iodine from a mixture with other solids.

The drying of solids

Solids are usually dried in a desiccator. The solid is placed in a container standing on a wire gauze over a drying agent such as anhydrous calcium chloride, silica gel or calcium oxide, and the vessel is sealed from the air.

Purification methods are summarized in Table 2.4.

Table 2.4 *Summary of purification methods*

Technique	What it separates
Solution and filtration	An insoluble substance from a soluble one
Crystallization	A crystalline solid from its solution
Separating funnel	Immiscible liquids
Simple distillation	A liquid from a solution, or two liquids with widely differing boiling points
Fractional distillation	Liquids with boiling points close together, and a better separation of other liquid mixtures
Chromatography	Many substances, as long as the components are adsorbed to differing extents on columns or paper, and also differ in their solubilities in the solvent being used
Sublimation	Solids, one of which sublimes on heating
Use of a drying agent	Water from a (nearly dry) solid or organic liquid. Typical drying agents (calcium oxide, silica gel, anhydrous calcium chloride) are used in desiccators to dry solids. *Note:* Some of these reagents are also used for drying gases, but then there are individual problems as described in 5.3.

2.3 Criteria of purity

It is often important to know whether a purification technique has worked, i.e. whether a chemical is pure or not. If the substance is a solid and its identity is known, its purity can be ascertained by making an accurate determination of the melting point. A pure solid has a sharp and fixed melting point (which can be found in a book of

46 Chemistry

data) and impurities will always lower the melting point. Some syllabuses could require you to describe in detail an experiment to determine accurately the melting point of a solid, for example by using a capillary tube strapped to a thermometer (see Figure) or by

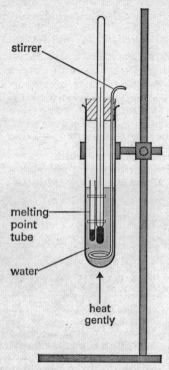

Melting point determination

working on a larger scale and plotting a cooling curve. If you use the latter, make sure that you understand why the horizontal part of the plot occurs at the melting point. The situation is very similar to that described at the boiling point (p. 16) where the temperature remains constant even though heat is still being applied. Here the change of state is from liquid to solid, and the temperature remains the same (even though heat is being *lost* to the atmosphere) as the inter-particle forces take over and the solid state is formed.

Boiling point determinations are used in a similar way in order to investigate the purity of a liquid (experimental details required).

Elements, compounds and mixtures. Purification 47

Impurities in a liquid make the boiling point increase. This method is not as reliable as a melting point determination as the boiling point is also affected by changes in atmospheric pressure.

Impurities make a melting point decrease.
Impurities make a boiling point increase.
Changes in atmospheric pressure also affect boiling points. (If pressure increases, boiling point increases, and vice versa.)

Questions

1 Four of the following substances have something in common. Which is the odd one?
(a) Nickel; (b) Copper; (c) Sulphur; (d) Silver; (e) Sodium.

2 Which of the following statements about salt water is FALSE?
   (a) It boils at a higher temperature than pure water.
   (b) It freezes at a higher temperature than pure water.
   (c) Its density is greater than that of pure water.
   (d) Its appearance is similar to that of pure water.

3 You are given a mixture of xylene and a dilute aqueous solution of potassium chloride. Xylene boils at 140 °C. Xylene is immiscible with water. Which of the following methods would you use to separate the mixture as far as possible?
(a) Filtration; (b) Use a separating funnel; (c) Distillation; (d) Chromatography; (e) Sublimation.

4 From the following methods of separating substances from each other, (a) filtration, (b) evaporation, (c) sublimation, (d) fractional distillation, (e) paper chromatography, select the most suitable for each of the following processes: (i) to separate a mixture of petrol and paraffin; (ii) to separate the various colours in red rose petals; (iii) to obtain salt from sea water.

5 The table below gives some details about the properties of three compounds.
   Use this information to devise a scheme for obtaining pure dry samples of the compounds from a mixture of the three, and carefully describe the procedure.
   Give one property of calcium fluoride in which it differs from calcium chloride.
   How could you prepare a dry sample of calcium fluoride from potassium fluoride solution?

Compound	Heat	Cold water	Hot water
Naphthalene	Sublimes	Insoluble	Insoluble
Calcium fluoride	No effect	Insoluble	Insoluble
Potassium chloride	No effect	Fairly soluble	More soluble

(JMB)

6 (a) It is required to separate a mixture of *three* solid dyes; one red, one yellow and one blue. The following facts are known about the dyes. The blue and yellow dyes are soluble in cold water, while the red dye is insoluble. When an excess of aluminium oxide is added to a stirred, green aqueous solution of the mixed blue and yellow dyes and the aluminium oxide is filtered off and washed with water. it is found that the solid residue is yellow and the filtrate blue. When the yellow solid is stirred with ethyl alcohol and the mixture filtered, the solid residue is white and the filtrate yellow.
   Describe how you would obtain dry samples of the three dyes.

(b) A green dye is known to be a hydrate. Describe how you would determine the percentage of water of crystallization in the hydrate. (JMB)

7 Describe briefly how you would separate a pure sample of the first-named substance

from the impurity in each of the following mixtures: (a) iron turnings contaminated with oil; (b) sodium chloride crystals contaminated with glass; (c) hydrogen sulphide contaminated with hydrogen chloride; (d) water contaminated with copper(II) sulphate); (e) copper powder contaminated with magnesium powder. (JMB)

8 A liquid Z was found to have a boiling point of 348 K (75 °C) at 760 mm pressure, and of 358 K (85 °C) at 800 mm pressure. At a pressure of 820 mm the boiling point of Z is likely to be: (a) 359 K (86 °C); (b) 361 K (88 °C); (c) 363 K (90 °C); (d) 365 K (92 °C); (e) 367 K (94 °C).

9 Give three reasons in each case why (a) air is considered to be a *mixture* of nitrogen and oxygen, (b) water is considered to be a *compound* of hydrogen and oxygen.

Draw a diagram of the apparatus you would use to obtain a sample of the air dissolved in tap water. How would you determine the proportion of oxygen in the air so obtained? How and why would your result differ from the proportion of oxygen in ordinary air? (JMB)

3 Electrochemistry. The electrochemical series, oxidation and reduction

3.1 The conduction of electricity by chemicals

The only substances which conduct electricity in the solid state are the metallic elements (which also conduct when molten) and the non-metallic element carbon (in the form of graphite). The only *compounds* which conduct electricity are those which are ionically bonded, and these only conduct when dissolved in water or when molten. Solutions or melts which contain only molecules (i.e. covalently bonded materials) will not conduct.

*An **electrolyte** is a solution or melt which conducts an electric current and is decomposed by it.*
***Electrolysis** is the chemical change which takes place when an electric current passes through an electrolyte.*
*An **electrode** is the metal or carbon rod by which current enters or leaves an electrolyte.*
*The **anode** (positive) is the electrode from which electrons leave the electrolyte.*
*The **cathode** (negative) is the electrode at which electrons enter the electrolyte.*
*A **cation** is a positively charged ion which will thus be attracted towards the cathode during electrolysis.*
*An **anion** is a negatively charged ion which will thus be attracted towards the anode during electrolysis.*

There are chemical changes when electrolytes conduct, but not when elements do so. Another difference in the way that elements and electrolytes conduct is the nature of the current passing through the chemical. An electric current is a flow of charged particles; in a metallic conductor it is a flow of electrons, but in electrolytes a flow of ions.

Elements	Compounds
Only the metals (solid or melted) and graphite conduct	Only ionic compounds conduct, and then only when melted or dissolved, never when solid
No chemical change when conduction takes place	Always change chemically when they conduct
Current is a flow of electrons	Current is a flow of ions

What happens during electrolysis?

You should be able to describe the details of an experiment which shows that an ionic solid (e.g. lead(II) bromide) becomes an electrolyte when molten. Solid ionic compounds cannot conduct because the ions are then not free to move from place to place. When molten or dissolved, ions can move towards the electrodes.

The process is easy to understand if we consider a molten compound such as lead(II) bromide, which will contain only two types of ion (Pb^{2+} and Br^-). Electrons from the battery go to the cathode and make it negatively charged. This attracts the positive ions (cations) to the cathode. Positive ions are atoms which have lost electrons, i.e. they have a 'vacancy' for electrons. The positive ions each take one, two or three electrons from the cathode until they become neutral atoms. In molten lead bromide, the cations are Pb^{2+} and on reaching the cathode these each receive two electrons and become lead atoms. Electrons are thus constantly removed from the cathode, and the neutral atoms formed join with others of their own kind to form stable structures. (Remember that individual atoms occur only in the noble gases; in this example, lead atoms join to form a giant structure, i.e. metallic lead.) A chemical change thus occurs at the cathode.

Negative ions (anions) are atoms with 'extra' electrons, and are attracted to the anode where they give up the 'extra' electrons to the anode and become neutral atoms. In this example, the Br^- ions each lose an electron to form bromine atoms. The atoms formed join together in pairs to form bromine molecules, Br_2. A chemical change thus also occurs at the anode.

The overall effect is that electrons are continually removed from the cathode and an equal number are fed in at the anode. Although electrons do not actually pass through the liquid, they *appear* to do so and a complete circuit is formed. The ions act as electron carriers, and are chemically changed in the process. This is summarized by ionic equations, e.g.

at the cathode $\quad Pb^{2+}(l) + 2e \rightarrow Pb(l)$
at the anode $\quad 2Br^-(l) - 2e \rightarrow Br_2(l)$

Electrochemistry. The electrochemical series, oxidation and reduction 51

The part water plays in electrolysis

Water is ionized only to a very small extent, but it plays an important part in the electrolysis of aqueous solutions.

$$H_2O(l) \rightleftharpoons H^+(aq) + OH^-(aq)$$

The reversible sign is important, for although only a very small proportion of molecules are ionized at any moment, as soon as any ions are removed then more are immediately formed (the Principle of Le Chatelier, p. 200). Strictly, the H^+ ion is too small to exist on its own and it joins with another water molecule to form the hydronium ion (oxonium ion) H_3O^+. We should thus write

$$H_2O(l) + H_2O(l) \rightarrow H_3O^+(aq) + OH^-(aq)$$

for the ionization of water. The simplified version will be used in this book; you should use the one appropriate for your syllabus.

When an electrolyte is dissolved in water there are usually four different kinds of ion present, i.e. those from the ionic compound and those from water. The events described earlier still occur at the electrodes, but only *one* kind of ion is discharged at each electrode. We say that at each electrode one of the ions is **selectively discharged**.

Note: if OH^- ions (e.g. from water) are discharged at the anode, they *always* form water and oxygen:

$$4OH^-(aq) - 4e \rightarrow 2H_2O(l) + O_2(g)$$

If H^+ ions are discharged, they always form hydrogen gas:

$$2H^+(aq) + 2e \rightarrow H_2(g)$$

Electrolysis experiments

If you are asked to describe a full electrolysis experiment, use the following headings.

1 Apparatus A fully labelled diagram is usually needed. There are several different types of voltameter or electrolysis cell and you must choose one appropriate for the chemical being electrolysed, e.g. if gases are discharged, choose an apparatus with which gases can be conveniently collected. Name the materials used for the electrodes, and label anode and cathode, adding (+) and (−) signs.

2 Ions present Show how the chemicals dissociate. The ions will come from one chemical if molten, and from two (including water) if in solution. For copper(II) sulphate solution this would be shown:

from copper(II) sulphate $\quad CuSO_4(aq) \rightarrow Cu^{2+}(aq) + SO_4^{2-}(aq)$
from water $\quad\quad\quad\quad\quad\quad\;\; H_2O(l) \rightarrow H^+(aq) + OH^-(aq)$

3 At the cathode State which ions are attracted to the cathode, which of them (if there is more than one type) is preferentially discharged, and show by means of an ionic equation what happens at the electrode.

4 At the anode Repeat (3) but with respect to the anode. Check at this stage that the final products in each case are capable of independent existence, e.g. Br_2 (bromine molecules) but not Br (bromine atoms).

5 Changes in the electrolyte It is sometimes useful to point out how the electrolyte changes during the process. For example, if sodium chloride solution is electrolysed, $Cl^-(aq)$ and $H^+(aq)$ ions are discharged so that the $Na^+(aq)$ and $OH^-(aq)$ ions also present become predominant in solution and sodium hydroxide solution is eventually formed from the original sodium chloride solution. Similarly, those aqueous solutions which produce hydrogen at the cathode and oxygen at the anode become more concentrated during the process because in effect they are losing water.

Specific examples of electrolysis

You will need experimental details of some of the examples summarized in Table 3.1, according to your syllabus. Note that the fifth example in the table is a special case. When electrodes are made of platinum or carbon they cannot themselves take part in any chemical changes as they are inert under these conditions. When a copper anode is used with a copper(II) salt solution there is, however, a third way in which the anode can receive electrons (i.e. in addition to the possibility of electrons from the two different anions in the solution). The atoms in the anode release electrons in forming copper ions, and these electrons pass on to the battery. The anode is thus seen to dissolve gradually.

Industrial uses of electrolysis
The extraction of reactive metals

Atoms of reactive metals readily lose electrons to form positive ions and they occur in nature as ions. The reverse process (i.e. the conversion back to metal atoms) is quite difficult, and can usually be achieved only by electrolysis. The essential steps in the production of aluminium and sodium are described below.

The extraction of aluminium

The ore is bauxite, impure aluminium oxide, Al_2O_3. This is purified by treatment with sodium hydroxide solution which dissolves the (amphoteric) aluminium oxide but not the impurities, which are

Table 3.1 Some examples of electrolysis

Ions present	At the cathode	At the anode
Copper(II) sulphate solution with platinum or carbon electrodes $CuSO_4 \rightarrow Cu^{2+}(aq) + SO_4^{2-}(aq)$ $H_2O \rightarrow H^+(aq) + OH^-(aq)$ Solution slowly becomes colourless dilute sulphuric acid.	Copper ions selectively discharged $2Cu^{2+}(aq) + 4e \rightarrow 2Cu(s)$	Hydroxide ions selectively discharged $4OH^-(aq) - 4e \rightarrow 2H_2O(l) + O_2(g)$
Sodium chloride solution with carbon electrodes $NaCl \rightarrow Na^+(aq) + Cl^-(aq)$ $H_2O \rightarrow H^+(aq) + OH^-(aq)$ Solution gradually becomes dilute sodium hydroxide solution.	Hydrogen ions selectively discharged $2H^+(aq) + 2e \rightarrow H_2(g)$	Chloride ions selectively discharged $2Cl^-(aq) - 2e \rightarrow Cl_2(g)$
Dilute sulphuric acid with platinum or carbon electrodes $H_2SO_4 \rightarrow 2H^+(aq) + SO_4^{2-}(aq)$ $H_2O \rightarrow H^+(aq) + OH^-(aq)$ The sulphuric acid slowly becomes more concentrated.	Hydrogen ions discharged $4H^+(aq) + 4e \rightarrow 2H_2(g)$	Hydroxide ions selectively discharged $4OH^-(aq) - 4e \rightarrow 2H_2O(l) + O_2(g)$
Sodium hydroxide solution with platinum or carbon electrodes $NaOH \rightarrow Na^+(aq) + OH^-(aq)$ $H_2O \rightarrow H^+(aq) + OH^-(aq)$ The sodium hydroxide solution slowly becomes more concentrated.	Hydrogen ions selectively discharged $4H^+(aq) + 4e \rightarrow 2H_2(g)$	Hydroxide ions discharged $4OH^-(aq) - 4e \rightarrow 2H_2O(l) + O_2(g)$
Copper(II) sulphate solution with copper electrodes $CuSO_4 \rightarrow Cu^{2+}(aq) + SO_4^{2-}(aq)$ $H_2O \rightarrow H^+(aq) + OH^-(aq)$	Copper ions selectively discharged $Cu^{2+}(aq) + 2e \rightarrow Cu(s)$	Copper atoms from the anode dissolve as ions $Cu(s) - 2e \rightarrow Cu^{2+}(aq)$

The anode loses mass and the cathode gains in mass by the same amount. The colour of the solution remains unchanged.

filtered off. Pure aluminium hydroxide is crystallized from the solution and then heated to produce the oxide.

$$2Al(OH)_3(s) \rightarrow Al_2O_3(s) + 3H_2O(g)$$

The oxide is dissolved in molten cryolite (Na_3AlF_6) and electrolysed in a large tank with carbon anodes and a carbon lining as the cathode (see Figure).

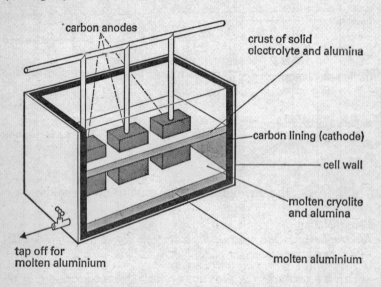

The extraction of aluminium by electrolysis

at the cathode $\quad 4Al^{3+}(l) + 12e \rightarrow 4Al(l)$
at the anode $\quad\;\; 6O^{2-}(l) - 12e \rightarrow 3O_2(g)$

The extraction of sodium

Sodium chloride (e.g. from rock salt) is mixed with a little calcium chloride and melted. The calcium chloride acts as an impurity and lowers the melting point of the sodium chloride, thus reducing energy requirements. The Down's cell (see Figure) uses a central carbon anode and a cylindrical iron cathode.

at the cathode $\quad 2Na^+(l) + 2e \rightarrow 2Na(l)$
at the anode $\quad\;\; 2Cl^-(l) - 2e \rightarrow Cl_2(g)$

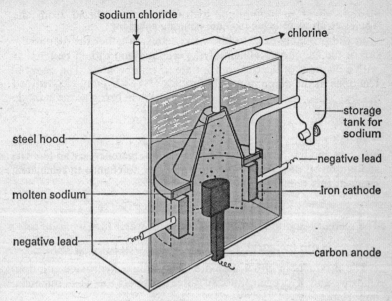

The Down's cell for the production of sodium

Electroplating
Electroplating is the process of coating an object (usually metallic) with a thin layer of another metal by electrolysis.

This is a useful way of protecting a metal from corrosion (e.g. by plating with nickel, chromium or gold) or of making a metal object more attractive.

The principle is to clean thoroughly the object to be plated, and to make it the cathode of a cell. The electrolyte is a solution of a salt of the metal which is to be plated on to the cathode. The anode is usually made of the same metal as that which is present in solution.

The purification of copper
This is basically a large scale version of the laboratory electrolysis of copper(II) sulphate solution using copper electrodes. The impure copper is made the anode and a piece of pure copper the cathode. Only pure copper atoms dissolve from the anode and are eventually deposited on the cathode, which increases in mass. This is an important process for copper is used extensively for electrical wiring and even slight traces of impurities make it a less efficient conductor.

The electrolysis of brine (sodium chloride solution)

You will probably need to know the details of either the diaphragm cell or the mercury (Kellner-solvay) cell, and you should be familiar with a diagram of the cell appropriate to your syllabus. The essential reactions in the diaphragm cell are

at the cathode $\quad 2H^+(aq) + 2e \to H_2(g)$
at the anode $\quad\quad 2Cl^-(aq) - 2e \to Cl_2(g)$

In the mercury cell the anode reaction is the same, but sodium ions are preferentially discharged at the mercury cathode. The sodium metal dissolves in mercury and has no chance to react with water at this point.

$$2Na^+(aq) + 2e \to 2Na(\text{in mercury})$$

The mercury–sodium amalgam reacts with water in a separate cell:

$$2Na(\text{in mercury}) + 2H_2O(l) \to 2NaOH(aq) + H_2(g)$$

The products in either case are thus sodium hydroxide solution, chlorine and hydrogen, and all three products are of commercial importance.

3.2 The electrochemical series (activity series)

If two different metals (or a metal and carbon) are dipped (as electrodes) into an electrolyte, and connected through a voltmeter, a voltage can be measured between the electrodes and a current flows between them. This is the reverse of electrolysis, for there is no battery to supply the current; the chemicals themselves produce a flow of electrons.

The metals can be arranged in a series according to the voltage recorded between them and a fixed electrode, and a typical list is shown in Table 3.2. In order to compare elements under carefully defined conditions the fixed electrode is the standard hydrogen electrode, and all other metals are connected to this in turn. The voltages produced when this is done under carefully controlled conditions are known as **electrode potentials**. Thus when a standard zinc electrode is connected to a standard hydrogen electrode, a voltage of 0.76 volts is recorded; this is said to be the electrode potential of zinc.

This order of electrode potentials for metals is also an order of

Electrochemistry. The electrochemical series, oxidation and reduction 57

Table 3.2 *Standard electrode potentials*

	Element	Standard electrode potential (volts)
Increasing chemical reactivity (For metals) ↑	sodium	−2.71
	magnesium	−2.36
	aluminium	−1.67
	zinc	−0.76
	iron	−0.44
	lead	−0.12
	hydrogen	0.00
	copper	+0.34
	mercury	+0.78
	silver	+0.79
	gold	+1.50

Important: The oxide layer on aluminium is very difficult to penetrate, and this often 'hides' the true chemical activity of the metal. In practice, the metal is often less reactive than zinc and iron because of this effect.

chemical activity, for a metal like sodium is chemically reactive because it readily loses electrons to another substance, and it produces a high electrode potential for the same reason. A negative electrode potential indicates that the substance tends to *lose* electrons to a hydrogen electrode (the more negative the potential, the greater is the ability to lose electrons), and a positive one means that the substance tends to *gain* electrons from a hydrogen electrode. Unreactive metals have positive electrode potentials.

A dry cell works on the principle that zinc (the outer case of the battery) and carbon (the central rod, capped with brass) are in contact with an electrolyte and will produce a current when joined externally.

The chemical activity of the elements follows the same trend as the electrode potentials, as the following examples show. This arrangement of elements in a list according to their chemical reactivity is called an **activity series**. Note that calcium is included in the examples, but not in Table 3.2; this is because the electrode potential of calcium is not as expected from its chemical activity, for reasons which are not normally explained at an elementary level. You may need to know experimental details of these illustrations of the activity series.

The reactions of the metals with water or steam

Table 3.3 *The reactions of the metals with water or steam*

	Metal	Reaction
↑ *Increasing chemical reactivity*	Sodium	Reacts vigorously with *cold* water $2Na(s) + 2H_2O(l) \rightarrow 2NaOH(aq) + H_2(g)$
	Calcium	Fairly vigorous reaction in *cold* water $Ca(s) + 2H_2O(l) \rightarrow Ca(OH)_2(aq) + H_2(g)$
	Magnesium	*Very* slow in cold water, rapid when heated in steam $Mg(s) + H_2O(l) \rightarrow MgO(s) + H_2(g)$
	Aluminium	No reaction (water or steam) due to protective oxide layer
	Zinc	No reaction in the cold, fairly rapid when heated in steam $Zn(s) + H_2O(g) \rightarrow ZnO(s) + H_2(g)$
	Iron	Reacts when heated in steam, reaction reversible $3Fe(s) + 4H_2O(g) \rightleftharpoons Fe_3O_4(s) + H_2(g)$
	Lead Copper Silver Gold	No reaction in water or steam

The reactions of the metals with oxygen in the air

Table 3.4 *The reactions of the metals with oxygen in the air*

	Metal	Reaction	Effect of water on oxide
↑ *Increasing chemical reactivity*	Sodium	Metal instantly tarnishes, i.e. forms oxide layer in cold. Burns readily when heated, with yellow flame. Off-white solid formed $4Na(s) + O_2(g) \rightarrow 2Na_2O$	Oxide reacts with water to form strongly alkaline solution
	Calcium	Rapidly covered with oxide layer in the cold. Sometimes difficult to ignite because of this. Fresh specimens burn readily when heated, with brick red flame. White solid formed $2Ca(s) + O_2(g) \rightarrow 2CaO(s)$	Oxide reacts with water to form strongly alkaline solution
	Magnesium	Oxide layer formed slowly when cold. Burns easily when heated, with dazzling white flame. White solid formed $2Mg(s) + O_2(g) \rightarrow 2MgO(s)$	Oxide is basic and reacts slightly to form weakly alkaline solution
	Aluminium	Oxide layer formed instantly in cold. Layer difficult to penetrate – protects metal beneath, which does not burn even when heated strongly $4Al(s) + 3O_2(g) \rightarrow 2Al_2O_3(s)$	Oxide insoluble in water. Amphoteric. May affect indicator due to impurities

Electrochemistry. The electrochemical series, oxidation and reduction 59

Table 3.4—*Contd.*

	Metal	Reaction	Effect of water on oxide
↑	Zinc	Oxide layer formed in cold. Turnings and pieces will not burn when heated, but the powder burns when heated, with blue-white flame. Oxide yellow when hot, white when cold $2Zn(s) + O_2(g) \rightarrow 2ZnO(s)$	Oxide insoluble in water. Amphoteric. May affect indicators due to impurities
Increasing chemical reactivity	Iron	Reacts in cold with water *and* oxygen (i.e. rusts) so not really comparable with the others. Heated *powder* sparkles as it oxidizes but does not really burn $3Fe(s) + 2O_2(g) \rightarrow Fe_3O_4(s)$	Basic oxide, insoluble in water, so no effect on indicators
	Lead	Oxide layer formed in the cold, but does not burn when heated $2Pb(s) + O_2(g) \rightarrow 2PbO(s)$	Amphoteric oxide, insoluble in water, so no effect on indicators
	Copper	Does not form any appreciable oxide layer in cold, although may react with gases in the air to form a green layer. Forms black oxide coating when heated, but this protects the metal below, which does not burn $2Cu(s) + O_2(g) \rightarrow 2CuO(s)$	Basic oxide, insoluble in water, so no effect on indicators
	Mercury	When heated forms red layer of mercury(II) oxide, does not burn $2Hg(l) + O_2(g) \rightarrow 2HgO(s)$	No effect on indicators
	Silver Gold	No oxide layer, even when heated	

Displacement reactions

If a metal is added to a solution of a salt of a metal which is lower in the activity series, it will displace this other metal from the solution.

For example, if zinc metal is added to copper(II) sulphate solution, zinc displaces the copper, i.e. copper metal is precipitated and the zinc dissolves as zinc ions. The solution thus slowly becomes zinc sulphate. These reactions are best shown by ionic equations, e.g.

$$Zn(s) + Cu^{2+}(aq) \rightarrow Zn^{2+}(aq) + Cu(s)$$

because the anion present (SO_4^{2-} in this case) is not directly involved in the reaction. In effect the two metals are competing for electrons; the better 'loser' of electrons (i.e. the more reactive metal) finishes as ions (having lost electrons) and the less reactive metal is changed from ions to metal atoms because it is made to gain electrons.

Note:
1 There will only be a reaction if the metal higher in the series starts as metal *atoms* (i.e. the free metal itself) *and* the other metal starts as *ions* (i.e. a solution of one of its salts).

2 Make sure that you can construct *balanced* ionic equations for these reactions, e.g. $Mg(s) + 2Ag^+(aq) \rightarrow Mg^{2+}(aq) + 2Ag(s)$.

3 If asked to *describe* any of these reactions, remember to include the colours and appearance of the metals which are precipitated, and also any colour changes which may occur in solution.

4 Aluminium does not take part in displacement reactions because of its oxide layer.

5 Do not quote sodium, potassium and calcium in displacement reactions because they also react with the water present.

Action of the metals on dilute acids

These are best regarded as displacement reactions. All those metals above hydrogen in the activity series will displace hydrogen from a solution of its ions, i.e. from dilute acids, which contain H^+ ions. Ionic equations can again be used, although the full equations are equally useful, e.g.

$$Zn(s) + 2H^+(aq) \rightarrow Zn^{2+}(aq) + H_2(g)$$
or, $$Zn(s) + 2HCl(aq) \rightarrow ZnCl_2(aq) + H_2(g)$$

Remember that nitric acid is *not* a typical acid in its reactions with metals, and hydrogen cannot be displaced from it in this way. Other points to note are included in Table 3.5.

Extraction of the metals

Metals such as gold which are low in the activity series, are sometimes found native (i.e. uncombined) in nature, and even if combined they are easily extracted from their compounds by chemical changes. This is a consequence of their unreactive nature, and these metals were the first to be found and used by man, e.g. copper and tin in the bronze age.

Metals high in the activity series are so reactive that they are always found combined (as ions), and such is their readiness to lose electrons that it is very difficult to change them from their ions back into the metals, i.e. their compounds are difficult to decompose. This is why sodium and aluminium, for example, were not discovered until comparatively recently. Electrolysis is normally used to make the ions of very reactive metals gain the electrons which convert them into metal atoms, e.g. the extraction of aluminium and sodium (p. 54). The chlorides are normally used for this purpose.

Metals which are intermediate in activity are often found as (or easily converted to) their oxides, which are reduced in a suitable furnace (e.g. the extraction of iron, p. 207).

Electrochemistry. The electrochemical series, oxidation and reduction 61

Table 3.5 *The reaction of metals with typical dilute acids*

Metal	Reaction with dilute hydrochloric acid	Reaction with dilute sulphuric acid
Sodium / Potassium	Too reactive to be used with any acids	
Calcium	Rapid effervescence in the cold to form hydrogen	Rapid effervescence initially in the cold, slowing down as insoluble calcium sulphate coats the metal
Magnesium	Rapid effervescence in the cold to form hydrogen	As with hydrochloric
Aluminium	Slow reaction in cold until oxide layer is penetrated; rapid if warmed, hydrogen formed	Very little reaction, oxide layer unbroken
Zinc	Steady effervescence to liberate hydrogen in the cold	Usually slow effervescence to form hydrogen unless the metal is impure
Iron	Fairly slow effervescence in the cold to give hydrogen	As with hydrochloric
Lead	No reaction – it is only just above hydrogen in the activity series	As with hydrochloric
Copper / Mercury / Silver / Gold	No reaction with these dilute acids	

(Increasing chemical reactivity ↑)

Typical equations:
$Mg(s) + 2HCl(aq) \rightarrow MgCl_2(aq) + H_2(g)$
$Mg(s) + H_2SO_4(aq) \rightarrow MgSO_4(aq) + H_2(g)$
$Zn(s) + 2HCl(aq) \rightarrow ZnCl_2(aq) + H_2(g)$
$Fe(s) + H_2SO_4(aq) \rightarrow FeSO_4(aq) + H_2(g)$

3.3 Oxidation and reduction

*A substance is said to be **oxidized** if it gains oxygen, loses hydrogen, or loses electrons. Such a reaction is an oxidation.*

*A substance is said to be **reduced** if it loses oxygen, gains hydrogen, or gains electrons. Such a reaction is a reduction.*

*A substance which brings about an oxidation is an **oxidizing agent**, and it is itself reduced during the reaction.*

*A substance which brings about a reduction is a **reducing agent**, and it is itself oxidized during the reaction.*

Note: an oxidizing agent must be itself reduced in performing its task, for if it (for example) removes electrons from something (i.e. oxidizes it), it must itself *receive* those electrons, i.e. be reduced. Think about this carefully; it is frequently misunderstood. As **red**uction normally occurs with **ox**idation, these reactions are often called **redox** reactions.

It is useful to think of electrolysis and the activity series in connection with oxidation and reduction.

All electrolytic changes at a cathode must be reductions as ions gain electrons.

All electrolytic changes at an anode must be oxidations as ions or atoms lose electrons.

Hydrogen is more likely to reduce oxides of metals low in the activity series (i.e. below hydrogen itself) rather than oxides of metals high in the series.

A metal will reduce an oxide of a metal lower than itself in the activity series.

Note: Think of these last two points as 'battles for oxygen'; the more reactive element wins.

There is no point in learning countless examples of oxidations and reductions, for a large proportion of all chemical reactions involve oxidation and reduction. It is more important that you should be able to recognize what is happening in a *given* reaction in terms of oxidation and reduction. This point will become more obvious if you look at typical examination questions on oxidation and reduction, such as the one given as a model answer on page 223.

It is also useful to be familiar with the common oxidizing and reducing agents and what they are capable of doing. The common agents are described in Tables 3.6 and 3.7, and typical uses are described in the appropriate Units, e.g. the conversion of iron(II)

Electrochemistry. The electrochemical series, oxidation and reduction 63

Table 3.6 *Some common reducing agents*

Reagent	Formula	Usually changes to	External signs of its action
Hydrogen	H_2	Water, by removing oxygen	Depends on the other reagent
Hydrogen sulphide	H_2S	Sulphur	Yellow solid formed
Moist sulphur dioxide or its solution	SO_2 or H_2SO_3	Sulphate ions	Depends on the other reagent
Reactive metals	—	Metal ions (by losing electrons) or metal oxide (by gaining oxygen)	Depends on the other reagent
Carbon	C	Carbon monoxide, which often burns to form carbon dioxide	Depends on the other reagent
Carbon monoxide	CO	Carbon dioxide by removing oxygen	Depends on the other reagent

Table 3.7 *Some common oxidizing agents*

Reagent	Formula	Usually changes to	External signs of its action
Acidified (H_2SO_4) potassium manganate(VII)	$KMnO_4$	Colourless Mn^{2+}(aq)	Colour change, purple to colourless
Acidified (H_2SO_4) potassium dichromate(VI)	$K_2Cr_2O_7$	Green Cr^{3+}(aq)	Colour change, orange to green
Oxygen	O_2	An oxide	Depends on the other reagent
Chlorine	Cl_2	A chloride or chloride ions	Depends on the other reagent
Concentrated sulphuric acid	H_2SO_4	Sulphur dioxide	Gas produced (sulphur dioxide) with characteristic smell
Concentrated nitric acid	HNO_3	Nitrogen dioxide	Brown gas evolved
Acidified hydrogen peroxide	H_2O_2	Water	Depends on the other reagent

salts to iron(III) salts on page 129. It is important to realize that the substances listed in Tables 3.6 and 3.7 are referred to as reducing agents and oxidizing agents not because they invariably act in that way, but simply because they commonly do so. A reducing agent can be made to act as an oxidizing agent if it reacts with a reducing agent stronger than itself, and the reverse is true of an oxidizing agent.

Remember the following points when answering questions on redox reactions.

1 If something is oxidized or reduced in a full chemical reaction, represented by a complete equation, there must also be a chemical which is reduced or oxidized. A reaction such as the one represented by the equation

$$Zn(s) \rightarrow Zn^{2+}(aq) + 2e$$

is *not* a complete reaction, as it involves only one chemical; it is a half reaction, and the equation is known as an ion-electron half equation.

2 In describing a reaction in terms of oxidation and reduction, always use expressions such as 'X is oxidized to Y by loss of electrons and W is reduced to Z by gaining electrons' rather than the temptingly simple 'X is oxidized and Y is reduced'.

3 Describe what you would *see* as evidence of oxidation and reduction if this is relevant to the question, e.g. colourless X is oxidized to yellow Y by ...'.

4 Electron transfer reactions are sometimes difficult to detect because electrons are not always shown in the equation. Thus in the reaction

$$2FeCl_2(aq) + Cl_2(g) \rightarrow 2FeCl_3(aq),$$

there appears to be no transfer of electrons, oxygen, or hydrogen. Nevertheless, oxidation and reduction are occurring because of electron transfer. Always look for atoms becoming ions or vice versa, or for an ion changing 'valency', for if either of these happens there must be a transfer of electrons. In the example above, iron(II) ions are changing to iron(III) ions and are thus oxidized by loss of electrons. At the same time some chlorine atoms are becoming chlorine ions, and they are being reduced by gaining electrons.

Electrochemistry. The electrochemical series, oxidation and reduction 65

Questions

1 An electric current was passed through an unknown solution. The gases which were evolved were collected and tested. The gas from the anode bleached damp litmus paper and the gas from the cathode burned with a squeaky pop. The solution was probably that of: (a) sodium hydroxide; (b) hydrochloric acid; (c) nitric acid; (d) sulphuric acid; (e) copper(II) sulphate.

2 Which of the following statements is the best definition of a cathode?
(a) It is the negatively charged electrode.
(b) It is the electrode at which electrons leave the electrolyte.
(c) It is the positively charged electrode.
(d) It is the electrode at which hydrogen is evolved.
(e) It is the electrode at which oxygen is evolved.

3 An electric current is passed through a solution of copper(II) sulphate, using platinum electrodes. The substance liberated at the anode is: (a) copper; (b) sulphur; (c) oxygen; (d) hydrogen; (e) sulphate.

4 Say whether the following are true or false:
(a) Copper will replace zinc from a solution of zinc nitrate.
(b) Iron will replace copper from a solution of copper(II) sulphate.
(c) Aluminium will replace magnesium from a solution of magnesium chloride.
(d) Iron will liberate hydrogen from dilute hydrochloric acid.
(e) Copper will liberate hydrogen from dilute sulphuric acid.

5 Write ionic equations for the following: (a) the reaction between zinc and copper(II) sulphate solution; (b) the reaction between magnesium and dilute hydrochloric acid; (c) the reaction at the cathode when copper is deposited; (d) the reaction at the anode when chlorine ions are discharged.
From these reactions, select two substances which have been oxidized and two which have been reduced, giving reasons for your answers.
Give one example of a 'redox reaction' and explain what it means.

6 Describe two experiments (one in each case) which you could use to demonstrate that zinc is above copper but below magnesium in the electrochemical series.

7 The following is a list of symbols of some of the elements in order of an 'activity series': K, Mg, Al, Zn, Fe, H, Cu, Ag.
(a) Which of these elements will not displace hydrogen from a dilute acid?
(b) Which of these elements has the most stable hydroxide?
(c) A piece of zinc is placed in iron(II) sulphate solution and a piece of iron is placed in zinc sulphate solution. In which solution would there be a reaction and why? Give the equation for the reaction.
(d) From these elements name (i) a metal which reacts with cold water; (ii) a different metal which reacts with hot water but only very slowly with cold; (iii) any other metals which will react when heated in a current of steam.
(e) Name any metals in the list whose heated oxides can be reduced by hydrogen. For one of these metals give an equation for the reaction.
(f) If mixtures of aluminium oxide and iron, and of iron oxide and aluminium, are heated, in which mixture is there a reaction and why? Give the equation for the reaction.
Outline any one experiment by which you could prepare dry crystals of zinc sulphate from a different zinc compound. (AEB)

8 Draw a diagram of an apparatus suitable for the electrolysis of copper(II) sulphate solution using platinum electrodes and for the collection of the products. Give the names and polarities of the electrodes, the names of the products, and the equations for the electrode reactions.

After passing the current for, say, ten minutes, what would be the effect of reversing the current and passing it in the opposite direction for about twenty minutes?

In the example given above, electricity is used to bring about a chemical change. Describe a way in which a chemical change can be used to release electrical energy. Carefully indicate the reactions which occur and where they take place. (JMB)

9 Why is it that oxidation and reduction reactions occur together? Select *two* reducing agents and for *each* describe a different reaction illustrating the above statement.

Name an oxidizing gas (not oxygen nor ozone), an oxidizing liquid (not a solution) and an oxidizing solid, and in *each* case give a different substance that can be oxidized by them, writing an appropriate equation. (CAM)

10 Name the product of each of the following reactions, and in each case state whether the reactant named in italics undergoes oxidation or reduction: (*a*) *hydrogen* burning in oxygen; (*b*) *chlorine* reacting with iron(II) chloride; (*c*) *hydrogen* reacting with nitrogen in the presence of a catalyst. (JMB)

11 Which of the following statements describes oxidation?
(*a*) Addition of hydrogen to a compound.
(*b*) A gain of one or more electrons.
(*c*) An increase in valency (oxidation state) of a metal.
(*d*) A decrease in the number of negatively charged ions present in the formula of a compound.
(*e*) Removal of oxygen.

4 Acids, bases and salts

4.1 Acids and bases

Acids

The acids normally used in elementary work are the three mineral acids (sulphuric, H_2SO_4; hydrochloric, HCl; nitric, HNO_3) and the organic acid ethanoic (acetic) acid, CH_3COOH. As the properties of the dilute and concentrated forms of the same acid are often very different, it is important to specify which type of acid you are referring to. This is either done in words (in written statements) or by using state symbols in equations. Thus $H_2SO_4(aq)$ refers to the dilute acid and $H_2SO_4(l)$ to the pure acid. Similarly, remember that $HCl(g)$ refers to the *gas* hydrogen chloride whereas $HCl(aq)$ means dilute hydrochloric acid; students sometimes read $HCl(g)$ to mean hydrochloric acid and perhaps seriously misinterpret a question.

What makes an acid an acid?

Dilute acids have certain properties in common, but before these are considered it is important to understand what makes all acidic solutions behave in a similar way. The key to the behaviour of acidic solutions is the solvent in which the acid is dissolved. You have probably seen experiments in which solid acids (e.g. citric, tartaric) and acids dissolved in dry organic (molecular) liquids such as methylbenzene (toluene) show none of the usual properties of acidic solutions (affecting indicators, effervescing with magnesium ribbon, etc.) as long as the other reagents are dry. As soon as water is added, however, the substances give the usual properties of acidic solutions.

It can also be shown that pure liquid acids cannot be electrolysed, but as soon as water is added they become both typical acidic solutions and electrolytes. The addition of water forms ions, and these ions are responsible for the behaviour we associate with acidic solutions.

When water is added to an acid, the acid molecules react with water molecules to form hydrogen (hydronium) ions. Only those substances which can do this form acidic solutions, and all acidic solutions have similar properties because they contain hydrogen (hydronium) ions.

An acid is a substance which is capable of giving up a proton (hydrogen ion) to a solvent (usually water) molecule.

e.g. $HCl(g)$ + *water* → $H^+(aq)$ + $Cl^-(aq)$
     acid   *solvent*   *hydrogen ion*   *acid anion*
  proton donor

or, more correctly,

$$HCl(g) + H_2O(l) \rightarrow H_3O^+(aq) + Cl^-(aq)$$

This definition was introduced by Brönsted and Lowry, and this theory of acids is known as the **Brönsted–Lowry theory**.

Properties of acidic solutions

The effect of acids on indicators These colour changes are discussed with those given by alkalis on page 70.

The action of dilute acids on carbonates and hydrogencarbonates

*All dilute acids react with **any** carbonate or hydrogencarbonate (whether solid or in solution) to form a salt, carbon dioxide (effervescence is observed), and water.*

e.g. $CuCO_3(s) + 2HCl(aq) \rightarrow CuCl_2(aq) + CO_2(g) + H_2O(l)$

Note: (1) Make sure that you can construct balanced equations for the reactions between dilute acids and a variety of metal carbonates. If you prefer to use ionic equations, remember that just one ionic equation applies to any carbonate + acid reaction:

$$CO_3^{2-}(aq \text{ or } s) + 2H^+(aq) \rightarrow CO_2(g) + H_2O(l)$$

The other ions which are present (from the carbonate and acid) exist free in solution as 'spectator ions', but can be made to combine to form a solid salt by evaporating off the water, as in salt preparations. Similarly, for all hydrogencarbonates:

$$HCO_3^-(aq) + H^+(aq) \rightarrow CO_2(g) + H_2O(l).$$

(2) The reaction between calcium carbonate and dilute sulphuric acid slows down and then stops, because the salt formed (calcium sulphate) is insoluble in water and coats the calcium carbonate, preventing further attack by the acid.

Acids, bases and salts 69

The action of dilute acids on metals These reactions are more difficult to classify, for what happens depends upon the individual metal and the individual acid. The reactions between metals and the typical dilute acids hydrochloric and sulphuric are summarized in Table 3.5 (p. 61). Nitric acid behaves differently, and does not usually evolve hydrogen when added to a metal. Nitric acid does, however, dissolve even copper and lead (but not aluminium) to form a solution of the metal nitrate, liberating oxides of nitrogen such as the brown gas nitrogen dioxide, NO_2.

Individual acids can be recognized by the appropriate test for the anion they contain, i.e. hydrochloric acid will give the tests for an acid (e.g. action on indicators or a carbonate) *and* also gives the chloride test (p. 170).

Bases and alkalis

These terms are often confused, but they are really very simple to understand. Bases form a large group of chemicals, and the alkalis are simply a special kind of base.

A base is a substance which can neutralize an acid to form a salt.
It is easy to recognize a base, for all metal oxides and hydroxides studied in elementary chemistry can act as bases.
*An **alkali** is a base which dissolves in water.*

There are many bases (e.g. magnesium oxide, iron(II) hydroxide and any other metal oxide or hydroxide) but only a very small number of alkalis. As the alkalis are frequently used, you should learn their names and formulae: sodium hydroxide, NaOH and potassium hydroxide, KOH are true alkalis; calcium hydroxide, $Ca(OH)_2$ (the solid is commonly called *slaked lime* and the solution *lime water*) and ammonia solution, $NH_3(aq)$ (sometimes called *ammonium hydroxide*) can be regarded as alkalis.

On the Brönsted-Lowry theory a base is a substance which is capable of accepting a proton.

Metal oxides and hydroxides are bases because oxide ions and hydroxide ions can accept protons from an acid:

$$O^{2-}(s) + 2H^+(aq) \rightarrow H_2O(l)$$
$$OH^-(aq) + H^+(aq) \rightarrow H_2O(l)$$

Ammonia gas is a base for a similar reason:

$$NH_3(g) + H^+(aq) \rightarrow NH_4Cl$$
(which can be shown $NH_4^+ Cl^-$).

Properties of bases

The action of bases (and acids) on indicators Only *soluble* bases, i.e. the alkalis, can effect indicators, and all acids and alkalis give these colour changes providing that they are in solution.

Indicator	Colour in acid	Colour in alkali
litmus solution	red	blue
red litmus paper	stays red	blue
blue litmus paper	red	stays blue
methyl orange	red-pink	yellow
universal	red, orange or yellow (see p. 71)	blue or purple
phenolphthalein	colourless	scarlet

The action of bases on ammonium compounds

When any base or alkali is warmed with an ammonium compound, the gas ammonia is given off and can be detected by the test described on page 169.

e.g. $Cu(OH)_2(s) + 2NH_4Cl(s) \rightarrow CaCl_2(s) + 2H_2O(l) + 2NH_3(g)$

An ionic equation can be used for this same reaction in solution

$$OH^-(aq) + NH_4^+(aq) \rightarrow NH_3(g) + H_2O(l)$$

We can make use of this reaction in three ways: (*a*) the laboratory preparation of ammonia gas, (*b*) in discovering whether an 'unknown' substance is an ammonium compound, by warming it with a known base and detecting the ammonia released and (*c*) in discovering whether an unknown substance is a base, by warming it with a known ammonium compound.

The action of bases and alkalis on metals Only two alkalis (sodium hydroxide and potassium hydroxide) react with metals, and even then only with two common metals, aluminium and zinc. The alkali solution has to be warmed. Hydrogen is given off and the metal dissolves to form the aluminate or zincate. You are unlikely to be asked for the appropriate equations, and the reactions are not important in elementary chemistry.

Alkalis as sources of hydroxide ions Alkalis in solution contain free $OH^-(aq)$ ions, and they are frequently used to precipitate metal hydroxides (p. 117).

4.2 pH and neutralization. Salts

The pH scale

All acidic solutions contain $H^+(aq)$ ions, and all alkaline solutions contain $OH^-(aq)$ ions. The pH scale measures the concentration of $H^+(aq)$ ions present in a solution, and normally runs from pH 0 to pH 14. A solution of pH 1 has a very high concentration of $H^+(aq)$ ions, i.e. it is very acidic. A solution of pH 13 has a very low concentration of $H^+(aq)$ ions but a high concentration of $OH^-(aq)$ ions, i.e. it is strongly alkaline. It is useful to remember that the pH scale changes in units of powers of ten, i.e. a solution of pH 1 has a concentration of $H^+(aq)$ ions 10 times greater than that in a solution of pH 2, and 100 times greater than that in a solution of pH 3.

Universal indicator is useful for detecting the pH of a solution because it is a mixture of indicators and can change to several different colours, each corresponding to a particular pH.

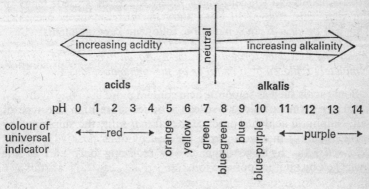

The pH scale

Neutralization

As acids and bases are chemical opposites, when they are mixed they 'destroy' (neutralize) each other. A neutral solution is neither acidic nor alkaline, and has a pH of 7.

If an acid is added slowly to an alkali, the pH of the solution gradually decreases (becomes less alkaline) as the acid neutralizes the alkali. If the addition of acid continues, after the neutralization of the alkali, the solution becomes acidic. The reverse process is possible when an alkali is added to an acid. When an *insoluble* base is added to an acid, the base neutralizes the acid as it is added, but when the acid has been completely neutralized the pH does not change (i.e. it stays at 7) no matter how much excess insoluble base is added, because only soluble bases (the alkalis) can affect pH. This important principle is used in the preparation of a salt by the addition of excess insoluble base to an acid (p. 73).

*You should be able to construct an equation for the neutralization of **any** acid by **any** base, for the products are **always** a salt and water only.*

Most neutralizations can also be expressed by an ionic equation, for they can be regarded as reactions between a provider of $H^+(aq)$ ions (an acid) and an acceptor of these ions (a base).

e.g. $$H^+(aq) + OH^-(aq) \rightarrow H_2O(l)$$

This equation is applicable to all neutralizations between acids and alkalis, for the other ions present do not join together unless the water is removed by evaporation, as in salt preparations.

Salts

Most students fail to realize how widely this term is used; many of the chemicals studied in elementary chemistry courses are salts. This will be clear from some of the following key statements.

*A **salt** is a substance formed when the hydrogen of an acid is partly or completely replaced by a metal or by the ammonium ion.*

Some acids such as sulphuric acid contain more than one hydrogen atom per formula unit and so can produce more than one type of salt. Sulphuric acid can produce the *normal salts*, the sulphates, by replacing all of the hydrogen in the acid by other positive ions, and *acid salts*, the hydrogensulphates, by replacing only part of the hydrogen by another positive ion.

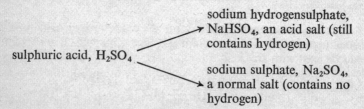

Hydrochloric acid forms salts called chlorides.
Nitric acid forms salts called nitrates.
Sulphuric acid forms salts called sulphates and hydrogensulphates.
Other less common acids form salts such as carbonates, hydrogencarbonates, iodides, bromides and sulphides.

All metallic (and ammonium) chlorides, sulphates, hydrogensulphates, nitrates, carbonates, hydrogencarbonates, iodides, bromides and sulphides are salts.

An acid which contains two hydrogen atoms replaceable by a metal (e.g. sulphuric acid) is said to be **dibasic**. An acid such as phosphoric(V) acid, H_3PO_4, is **tribasic**.

Methods of preparing salts

It is important to learn the *principle* of each method rather than specific examples, for by simply changing the names of the chemicals literally hundreds of different salts can be made.

1 The neutralization of a base by an acid

There are three ways in which this can be done, but all three have exactly the same reaction:

$$\text{acid} + \text{base} \to \text{a salt} + \text{water}$$

(a) Using an insoluble base and an acid This involves the addition of excess solid base (i.e. until no more will 'dissolve') to an acid so that the acid is completely neutralized, but the excess solid does not affect the pH of the solution and can be filtered off. (Experimental details required.)

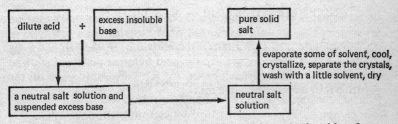

This is a very general method, for the oxide or hydroxide of any metal (except for those of sodium and potassium, which are soluble) can be used with any acid.

e.g. $CuO(s) + H_2SO_4(aq) \to CuSO_4(aq) + H_2O(l)$
$Mg(OH)_2(s) + 2HCl(aq) \to MgCl_2(aq) + 2H_2O(l)$

(b) Using a soluble base (alkali) and an acid to form a normal salt This involves the exact and complete neutralization of two solutions using a burette, pipette and an indicator. The alkali is nearly always sodium or potassium hydroxide, and the acid is usually one of the mineral acids. (Experimental details required.) A typical sequence would be:

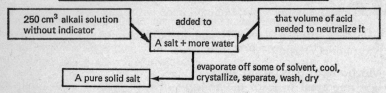

Typical salts which could be made by this method are KCl, NaCl, K_2SO_4, Na_2SO_4, $NaNO_3$ and KNO_3.

e.g. $\quad NaOH(aq) + HCl(aq) \rightarrow NaCl(aq) + H_2O(l)$

(c) Using a soluble base (alkali) and an acid to form an acid salt This is less common than (a) and (b), and is not required in some syllabuses. The only salts likely to be made in this way are sodium and potassium hydrogensulphates. The procedure is similar to that in (b), and the volume of alkali required to completely neutralize a sample of sulphuric acid is determined. (The solution produced contains the normal salt.) The experiment is then repeated, using the same volume of acid but only half the volume of alkali needed for complete neutralization, when the acid salt is formed in solution.

e.g. $\quad 2NaOH(aq) + H_2SO_4(aq) \rightarrow Na_2SO_4(aq) + 2H_2O(l)$
$\quad\quad\quad\quad\quad\quad\quad\quad\quad\quad$ (normal salt)

$\quad\quad NaOH(aq) + H_2SO_4(aq) \rightarrow NaHSO_4(aq) + H_2O(l)$
$\quad\quad\quad\quad\quad\quad\quad\quad\quad\quad$ (acid salt)

2 The neutralization of an acid by a carbonate

Almost any combination of acid and carbonate will react to form a salt, carbon dioxide and water. Most metal carbonates (except those of sodium and potassium) are insoluble in water and so an excess can be added to ensure that the acid is completely neutralized. The excess solid does not affect the solution in any way, and is filtered off. (Experimental details required.)

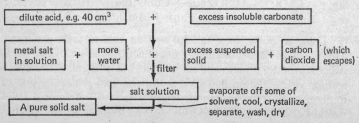

If sodium or potassium carbonates are used, only enough to neutralize the acid should be added, i.e. add in small quantities with stirring until addition no longer causes effervescence. Any excess would dissolve in the water and contaminate the salt.

3 The neutralization of an acid by a metal

Always think carefully before you describe a salt preparation involving this reaction, for not all combinations of acid and metal will react (p. 61). Nitric acid can be neutralized by *most* metals to form a solution of the nitrate, but this acid often needs to be warmed and the neutralization should be conducted in a fume cupboard as poisonous oxides of nitrogen will be produced. As with earlier methods, an *excess* of the solid is used to ensure that the acid is neutralized. The following sequence is typical of the preparation of a metal nitrate by this method. If hydrochloric or sulphuric acids are used the principle is the same, but fewer metals are suitable and the fume cupboard is not necessary.

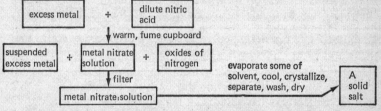

4 Precipitation reactions (double decomposition)

Salts which are insoluble in water cannot be made by methods 1 to 3, all of which involve the crystallization of a *soluble* salt from solution.

Relatively few insoluble salts are encountered at 'O' Level, and so these need to be known. The common ones are:
  chlorides: *AgCl, $PbCl_2$*
  sulphates: *$PbSO_4$, $BaSO_4$, $CaSO_4$ (slightly soluble)*
  carbonates: *all insoluble except those of sodium, potassium*
  nitrates: **no** *nitrates are insoluble*
    PbI_2 is also insoluble

The principle of the method is to add a **solution** *containing the required metal ion to a* **solution** *containing the required anion, filter or centrifuge off the precipitate, wash and dry it.*

Note that both starting materials must be soluble, so choose them carefully. As *all* metal nitrates are soluble, you can guarantee that the appropriate metal nitrate will be a suitable solution to provide the metal ion. Similarly, as *all* sodium salts are soluble, a solution of the appropriate sodium compound can be used for the anion.

These reactions are best shown by ionic equations. The sodium ions and nitrate ions (if a sodium salt and a nitrate are used) are 'irrelevant' (spectator ions) and remain unchanged in solution.

e.g. $\quad Pb^{2+}(aq) \quad + \quad 2Cl^{-}(aq) \quad \rightarrow PbCl_2(s)$
$\quad\quad$ (from lead nitrate) $\quad$ (from sodium chloride)

5 Direct combination between elements

This is not a general or convenient method, but certain specific reactions of this type are important and are included in some syllabuses. Simple salts such as chlorides and sulphides, which contain only two elements, can be made by joining the elements together. The chlorides can be made by heating the element in chlorine, and you may have formed the salt iron(II) sulphide by warming a mixture of iron and sulphur.

This method is particularly important for the preparation of the *anhydrous* chlorides of iron(III) and aluminium. Their *hydrated* chlorides can be made easily by some of the methods already described, but the hydrated chloride cannot then be made anhydrous by heating, because the water of crystallization given off reacts with (hydrolyses) the salt. The anhydrous salt is made by passing dry chlorine over the heated metal in an apparatus designed to keep atmospheric moisture away from the product, which sublimes into a cool receiver. If you have studied a reaction of this type you should be able to describe the preparation and draw a diagram of the apparatus. A similar preparation is also needed for anhydrous iron(II) chloride, but this time hydrogen chloride gas is used instead of chlorine.

$$2Al(s) + 3Cl_2(g) \rightarrow 2AlCl_3(s)$$
$$2Fe(s) + 3Cl_2(g) \rightarrow 2FeCl_3(s)$$
$$Fe(s) + 2HCl(g) \rightarrow FeCl_2(s) + H_2(g)$$

Some important points about salt preparations

1 A particular soluble salt can usually be prepared by several methods, e.g. zinc chloride by zinc metal and hydrochloric acid (metal + acid), zinc oxide and hydrochloric acid (base + acid), zinc hydroxide and hydrochloric acid (base + acid) and zinc carbonate and hydrochloric acid (carbonate + acid). You will have to decide which of the methods is appropriate or convenient for a particular examination question.
2 Do not worry in an examination if you are asked to describe the preparation of a particular salt and you cannot recall ever having made it; you should use the principles given in this section to plan

Acids, bases and salts 77

a suitable experiment. Remember that you may not be told that a particular chemical is a salt – it is up to you to recognize a salt from its name or formula.

4.3 Strong and weak acids and bases

The reaction between an acid and its solvent is reversible, and the **strength** of the acid decides the extent to which the forward reaction proceeds. If the acid liberates protons readily and forms a high concentration of $H^+(aq)$ ions when added to water, it is a strong acid. A weak acid is hardly dissociated (split up into separate ions) when added to water; the equilibrium in this case is to the left and the concentration of $H^+(aq)$ ions is low.

To make this easier to understand, consider what might happen when, say, a hundred 'units' of (1) a strong acid and (2) a weak acid are allowed to reach equilibrium when dissolved in the same volume of water. HA is the formula of the acid in each case.

(1) *Strong acid*

	water + HA	$\rightleftharpoons$ $H^+(aq)$	+ $A^-(aq)$
Initial number of units	100	0	0
Units at equilibrium	1	99	99

(2) *Weak acid*

	water + HA	$\rightleftharpoons$ $H^+(aq)$	+ $A^-(aq)$
Initial number of units	100	0	0
Units at equilibrium	99	1	1

In this example, (1) is 99 times as strong as (2) because 99 times as many acid units, $H^+(aq)$, are produced from a given number of acid molecules.

The mineral acids are all strong acids, but ethanoic (acetic) acid, citric acid, methanoic (formic) acid and tartaric acid are examples of weak acids. Similarly, sodium hydroxide is a strong alkali because it dissociates almost completely when added to water to produce a high concentration of $OH^-(aq)$ ions; but ammonia solution is only a weak alkali.

It is important not to confuse the words 'concentrated' and 'strong' (or 'dilute' and 'weak'). These words have very special meanings in chemistry. *Concentrated* or *dilute* refers to the concentration of the acid, i.e. how much acid there is in a certain volume of water. *Strong* and *weak* refer to the degree of dissociation of the substance into ions. It is quite possible for a concentrated acid to be a very weak acid.

5 The air, gases, oxygen and hydrogen

5.1 The air

The air is a mixture, and its composition varies from time to time and from place to place. The following is a typical composition by volume of *dry* air. Ordinary air always contains water vapour, but the proportion of this varies more than any of the other components.

Component	Composition by volume (%)
nitrogen	78.08
oxygen	20.95
argon	0.93
carbon dioxide	0.03
neon	0.002
other noble gases	0.0006

The air also contains very small traces of methane and ozone, and solids such as soot, bacteria and pollen. In addition, it is likely to contain pollutant gases such as hydrogen sulphide and sulphur dioxide (especially over industrial areas) and carbon monoxide (particularly over towns and cities).

You could be asked to describe an experiment which shows that air contains an 'active' gas (oxygen) and that this gas occupies one fifth of the total volume of the air. A typical experiment to show this involves two syringes connected together by a combustion tube containing copper powder; you should revise the experimental details of whichever experiment you have used. Some simple chemical principles can be used to remove carbon dioxide, water vapour and oxygen from a sample of air so as to leave a relatively pure sample of nitrogen (but still containing traces of noble gases). If you have seen such an experiment, revise the details.

The solubility of air in water

Air contains water vapour, and water also contains dissolved air. The solubility of most solids in water increases as the temperature rises, but the opposite is true of the solubility of gases in water, so that dissolved air can be 'boiled out' of water. You could be asked to describe an experiment to show that water contains dissolved air.

If a sample of the air boiled out of water is analysed it is found that its composition is different from that of ordinary air. This is because the different components of air dissolve in water to different extents. Oxygen is more soluble in water than nitrogen, so air dissolved in water is richer in oxygen (about 35% by volume) than ordinary air.

5.2 Combustion, respiration, photosynthesis and air pollution

Combustion and respiration normally utilize oxygen (from the air), and air pollution often arises from combustion processes. Photosynthesis helps to balance these effects by putting oxygen back into the air. These processes thus have oxygen as a common factor.

Combustion

Combustion is the chemical combination of a substance with a gas (usually oxygen) with the liberation of heat and/or light.

The term can be applied to gases other than oxygen, for some elements burn in other gases, e.g. sodium burns in chlorine to form sodium chloride.

Various things can happen to chemicals when they are heated in air. Some substances decompose and lose mass (e.g. hydrated copper(II) sulphate crystals), some remain unchanged (e.g. silicon dioxide, encountered in nature as quartz and sand) and others gain in mass because they combine with the oxygen in the air. All combustions involve an increase in mass of the substance being burned, because it combines with oxygen to form one or more new products. If a substance such as candle-wax is burned, it appears to lose mass, but this is because the products of combustion are gases which escape; if these were collected and weighed, it would be found that the mass of the products of combustion is greater than the loss in mass of the candle.

The combustion of *elements* is considered in detail in 5.4, but many of the important combustion processes involve the burning of *compounds* in air. Most of our fuels are hydrocarbons (i.e. compounds of hydrogen and carbon only), and the complete combustion of these substances always produces carbon dioxide (from the carbon present), water vapour (from the hydrogen present) and energy (e.g. heat). This energy is used to heat our homes, produce electricity and so on. Typical hydrocarbons include petrol, paraffin and natural gas (methane), e.g.

$$CH_4(g) + 2O_2(g) \rightarrow CO_2(g) + 2H_2O(g)$$
$$C_5H_{12}(l) + 8O_2(g) \rightarrow 5CO_2(g) + 6H_2O(g)$$

Respiration

This is not the same as breathing. All living things respire, but do not all breathe.

Respiration is the process by which a living organism obtains its energy from food substances, and it can be considered as a type of combustion process.

Breathing is the process used by organisms to ensure that oxygen enters the body (so that respiration, i.e. the combustion of food, can take place) and to remove the products of combustion from the body.

The foods used by living organisms (e.g. carbohydrates and fats) always contain hydrogen and carbon, but they are not hydrocarbons because they also contain other elements such as oxygen. Respiration is essentially similar to combustion in that the carbon and the hydrogen in the food is oxidized by oxygen to carbon dioxide and water, and energy is released, but respiration occurs in a series of small steps and the food is not actually 'burned'. Equations such as the following summarize these many small steps:

$$C_6H_{12}O_6(aq) + 6O_2(g) \rightarrow 6CO_2(g) + 6H_2O(l) + \text{energy}$$
(a sugar)

Photosynthesis

Photosynthesis is the process by which green plants produce carbohydrate foods (sugars and starches) from carbon dioxide and water, with the liberation of oxygen. Chlorophyll (a green pigment) and energy from sunlight are needed for the process. Summary equation:

$$energy + 6CO_2(g) + 6H_2O(l) \xrightarrow{chlorophyll} C_6H_{12}O_6(aq) + 6O_2(g)$$

In effect this is the reverse of respiration. It is one of the most important chemical reactions in the world, for two reasons. All of the food we eat has its origin in photosynthesis, for even meat is produced by animals which have themselves eaten plant sugars and starches. It is also obvious that without photosynthesis our oxygen supply would have been exhausted by the processes of combustion and respiration.

Air pollution
The substances which pollute the air can be divided into four categories.

1 Toxic gases
These arise either because of incomplete combustion of fuels or because of the combustion of impurities in fuels.

fuels
- *incomplete combustion*, e.g. because of poor ventilation or because of blockages in apparatus or simply inefficient burning (e.g. badly ventilated gas fires, and also petrol and diesel engines) → carbon monoxide and even elemental carbon. Carbon monoxide is very dangerous – odourless, invisible, remains unchanged in the atmosphere and toxic
- *combustion of impurities in the fuel*, e.g. sulphur and nitrogen compounds → oxides of sulphur (SO_2 and SO_3), hydrogen sulphide (H_2S) and oxides of nitrogen (e.g. NO_2). These gases are toxic and corrosive

Steps are being taken to improve the combustion of petrol and oil in internal combustion engines, fuels are being processed more carefully to remove sulphur compounds and industry is constantly trying to improve its filter systems and other methods used to remove toxic gases from the waste products.

2 Solid particles in the air
The main pollutant of this type is carbon. Smokeless zones have been introduced to reduce the amount of carbon (e.g. as smoke) put into the air, for smoke (be it tobacco smoke or any other form) is detrimental to health because it often contains carcinogenic (cancer-producing) hydrocarbons, as well as carbon. Smokeless solid fuels have been produced so that all of the volatile material and potential smoke have been removed, and industry has had to use more efficient smoke and dust extractors.

3 Radioactive substances in the air
Nuclear power inevitably produces radioactive waste materials. These are not likely to affect the

atmosphere seriously, although there is a potential risk to the sea and the land. The explosion of any form of nuclear weapon does, however, add to the natural radioactivity in the atmosphere, and successful attempts are being made internationally to reduce or stop nuclear experiments of this type.

4 Heavy metals in the air Ions of heavy metals (e.g. lead and mercury) are usually toxic, and unfortunately tend to be accumulative poisons, i.e. they build up in the body and are not excreted. There is concern over the way in which these compounds are sometimes found in liquid industrial waste (which eventually pollutes the rivers and the seas) and they are also found in the air.

Lead occurs in the air as lead compounds formed by the combustion of petrol. Petrol contains several additives, one of which is tetraethyl lead(IV), $Pb(C_2H_5)_4$, which is important in preventing premature explosion of the petrol/air mixture in the cylinders of internal combustion engines.

5.3 Gas preparations

A gas is made in some form of generator, and leaves the generator by a delivery tube, which may lead straight to a collection vessel or via a purification (e.g. drying) stage. The method of collection depends upon the physical properties of the gas. Note particularly:

1 If the generator flask is to be heated, it must have a round bottom.
2 Never dry a gas and then collect it over water.
3 Never collect a gas over water if it is fairly soluble in water or reacts with it.
4 Make sure that when tubes deliver gas *into* a liquid, they dip below the liquid, and that those tubes which take a gas away from a liquid do not enter the liquid. Similarly, ensure that delivery tubes *enter* gas jars (if used) and that gas jars are the right way up for the gas being collected.
5 If a toxic or unpleasant gas is being prepared, state that a fume cupboard is needed.
6 Always think of the scale of operation – do not use a test tube as a generator if large quantities of gas are needed.
7 In any written description, always state that the first sample of gas to be collected is then discarded; it is mainly air.
8 When drawing a diagram, make sure that the different pieces of apparatus are drawn to the same scale! Too many students give the impression that they have never actually seen the apparatus they draw!

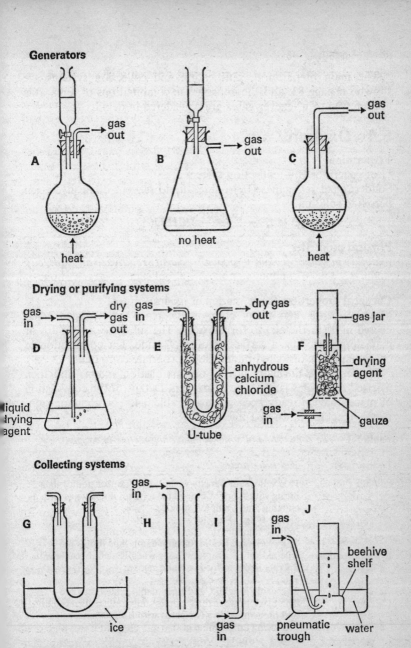

Apparatus used for preparing, purifying, and collecting gases

The main preparation, purification and collection systems are shown on page 83, and the appropriate combinations of apparatus are referred to whenever gases are considered in detail.

5.4 Oxygen

Preparation

Use: generator B + collection system J
Solid reagent manganese(IV) oxide, liquid reagent dilute hydrogen peroxide solution.

$$2H_2O_2(aq) \xrightarrow{\text{catalyst of } MnO_2} 2H_2O(l) + O_2(g)$$

Physical properties

solubility in water	*colour*	*odour*	*density relative to air*	*toxicity*
slight	none	none	about same	none

Chemical properties; the formation of oxides

Most elements, and many compounds, react with oxygen when heated in it (or the air) to form oxides. The substance may burn or may simply produce a surface coating of oxide. The combustion of compounds is considered under the appropriate compound.

Elements are normally heated on combustion spoons and then inserted into a gas jar or tube of oxygen. Details of the combustion of metals have already been given in Table 3.4. The common non-metals react as shown in Table 5.1.

Table 5.1 *The combustion of some common non-metals in oxygen*

Element	Reaction details
Sulphur	Pale yellow solid melts when heated (amber liquid) then burns with a bright blue flame to form a misty gas with a choking smell, sulphur dioxide. $S(s) + O_2(g) \rightarrow SO_2(g)$ (a little sulphur trioxide, SO_3, is also formed). The oxides are acidic, turning damp blue litmus red
Red phosphorus	The red powder rapidly reacts when heated, burning with a white flame to form a white smoke: $P_4(s) + 5O_2(g) \rightarrow P_4O_{10}(s)$ The oxide is acidic, and turns damp blue litmus red
Carbon	The black powder (or granules) only reacts when red hot, and then smoulders (or burns with a white flame) producing an invisible gas: $C(s) + O_2(g) \rightarrow CO_2(g)$ The gas formed is weakly acidic, turning damp blue litmus a purple-red colour

Oxides

1 *Most metal oxides have the properties of a base. Most non-metal oxides are acidic or neutral.*
2 *Some metal oxides have the properties of both an acid and a base and are said to be **amphoteric**, e.g. ZnO, Al_2O_3.*
3 *A neutral oxide is not simply an oxide which does not dissolve in water; it is also insoluble in both acidic and alkaline solutions.*
4 An oxide which contains two oxygen atoms joined directly together, i.e. a O—O link, is called a **peroxide**. There are only two common examples: hydrogen peroxide, H_2O_2, and sodium peroxide, Na_2O_2. Peroxides of metals are salts of H_2O_2, and contain the peroxide ion, O_2^{2-}.
5 A non-metallic oxide which dissolves in water to form an acid is also known as the **anhydride** of that acid, i.e. the acid without water. Sulphur trioxide is the acid anhydride of sulphuric acid ($SO_3(g) + H_2O(l) \rightarrow H_2SO_4(aq)$) and nitrogen dioxide is a mixed anhydride for it forms two acids in this way

$$2NO_2(g) + H_2O(l) \rightarrow HNO_3(aq) + HNO_2(aq).$$

The properties of the common metallic oxides are given in Table 8.2 (p. 116) and the properties of some common non-metallic oxides are given in Table 5.2.

Table 5.2 *Some common non-metallic oxides*

Oxide, formula and state at room temperature	Effect on water, and type of oxide
Sulphur dioxide, SO_2, gas	Some dissolves, some reacts: $SO_2(g) + H_2O(l) \rightarrow H_2SO_3(aq)$ Acidic oxide, the anhydride of sulphurous acid
Sulphur trioxide, SO_3, white solid usually seen as a smoke	Violent reaction: $SO_3(s) + H_2O(l) \rightarrow H_2SO_4(aq)$ Acidic oxide, the anhydride of sulphuric acid
Nitrogen dioxide, NO_2, gas	Reacts to form two acids: $2NO_2(g) + H_2O(l) \rightarrow HNO_3(aq) + HNO_2$ Acidic oxide, a mixed anhydride
Silicon dioxide, SiO_2, solid	Insoluble. Neutral to indicators, but is an acidic oxide which dissolves in hot alkali
Carbon dioxide, CO_2, gas	Some dissolves and some reacts: $CO_2(g) + H_2O(l) \rightarrow H_2CO_3(aq)$ Weakly acidic, the anhydride of the unstable carbonic acid
Carbon monoxide, CO, gas	Insoluble; neutral oxide
Water, H_2O, liquid	Neutral oxide
Phosphorus(V) oxide, P_4O_{10}, solid	Reacts violently to form a solution of phosphoric(V) acid, H_3PO_4

Uses of oxygen

1 When mixed with ethyne (acetylene), forms a mixture which burns at a very high temperature (oxy-acetylene flame) and is used in welding.
2 Steel making; vast quantities are used in this way.
3 In rocket fuels.
4 Respiratory aid (hospitals, climbing, diving, high altitude flying).

Test for oxygen
See page 169.

5.5 Hydrogen

Preparation
Use generator B + collection system J or I (see Figure, p. 83).
Solid reagent zinc metal (turnings), liquid reagent dilute hydrochloric acid (*not* sulphuric).

$$Zn(s) + 2HCl(aq) \rightarrow ZnCl_2(aq) + H_2(g).$$

No naked flames near the apparatus.

Physical properties

solubility in water	colour	odour	density relative to air	toxicity
slight	none	none	much less dense	none

Note: see also diffusion of hydrogen, p. 19.

Chemical properties
Combustion A *mixture* of hydrogen with air or oxygen explodes when sparked or ignited, and for this reason any attempt to burn hydrogen is potentially dangerous. On a small scale (test-tube) this explosion (pop) is used as a test for the gas.

If *pure* hydrogen is burned in air or oxygen, it burns with a blue flame and produces steam as the only product, which condenses to liquid water. You could be asked to describe an experiment which shows this, and if so you should draw attention to the precautions needed.

$$2H_2(g) + O_2(g) \rightarrow 2H_2O(g)$$

As a reducing agent Hydrogen will reduce the heated oxides of unreactive metals to the metal, itself being oxidized to steam (p. 62). You should be able to draw a typical apparatus used for these reductions, and (if relevant to your syllabus) understand that this kind of reaction can be used to determine the empirical formula of a compound (p. 105) and to verify the Law of Constant Composition and the Law of Multiple Proportions (p. 103).

Reaction with metals If a stream of hydrogen is passed over hot sodium metal a white crystalline solid, sodium hydride, is formed.

$$2Na(s) + H_2(g) \rightarrow 2NaH(s)$$

The compound is ionic and contains the relatively rare H^- ion. When this compound is molten and electrolysed, hydrogen is given off at the *anode*. Only the reactive metals (e.g. sodium, potassium and calcium) form hydrides which are ionic, and the compounds are all solids at room temperature.

Reaction with non-metals Hydrogen will react under appropriate conditions with most non-metals to form *covalent* hydrides, which are usually gases at room temperature (c.f. metal hydrides).

e.g.
$$F_2(g) + H_2(g) \xrightarrow{\text{explosive even in dark}} 2HF(g)$$
$$Cl_2(g) + H_2(g) \xrightarrow{\text{explosive in sunlight}} 2HCl(g)$$
$$2N_2(g) + 3H_2(g) \xrightarrow[\text{catalyst (Haber process)}]{\text{high temperature, pressure}} 2NH_3(g)$$

Uses of hydrogen
1 The manufacture of many organic chemicals, e.g. methanol and nylon.
2 The manufacture of ammonia by the Haber process.
3 The manufacture of cooking fats and margarine; heated liquid plant oils (e.g. corn oil) are heated with hydrogen under pressure in the presence of a nickel catalyst, and are converted to solid fats.

Questions

1 Which of the following does *not* reduce the amount of oxygen in the air?
   (a) Aerobic respiration.
   (b) Rusting.
   (c) Photosynthesis.
   (d) The use of petrol as a fuel.
   (e) The combustion of natural gas.

2 Which of the following statements about hydrogen is untrue?
   (a) It is a neutral gas, almost insoluble in water.
   (b) It is a reducing agent.
   (c) It will burn in air to form steam.
   (d) It diffuses more rapidly than carbon dioxide.
   (e) It is prepared by the action of dilute nitric acid on zinc.

88 Chemistry

3 Which of the following statements about oxygen is untrue?
  (a) It is a very electronegative element.
  (b) It can be prepared by the catalytic decomposition of hydrogen peroxide.
  (c) It forms acidic oxides with most metals.
  (d) It has no smell or taste.
  (e) It allows most substances to burn more vigorously than they do in air.

4 Draw a labelled diagram of the apparatus you would use to prepare and collect gas jars of *either* oxygen or hydrogen (not by electrolysis). Describe briefly *three* experiments you have seen that demonstrate chemical or physical properties of the gas that you have chosen above. In any chemical reaction mentioned, name and describe the product(s). Give two important different uses of *each* gas (excluding balloons). (CAM)

5 Describe an experiment that you could perform to find, as accurately as you can, the percentage by volume of oxygen in the air. Nitrogen as normally obtained from the air has a slightly greater density than nitrogen prepared from a compound. Give the reason for this greater density.

Give two natural ways by which nitrogen is returned to the soil. Describe what you would observe and say what is formed when: (a) solid sodium hydroxide is left in a dish and exposed to the air; (b) copper foil is heated in the air. (CAM)

6 Name the products formed when the following react with an excess of oxygen: (a) carbon; (b) magnesium; (c) hydrogen; (d) zinc.

Write equations for the reactions, if any, of these products with (i) dilute hydrochloric acid, (ii) sodium hydroxide solution. If no reaction occurs, write 'no reaction'.

State the type of oxide formed by each of the four elements. (JMB)

7 Compare and contrast the processes combustion, respiration and photosynthesis.

6 The Periodic Table. Structure

6.1 The Periodic Table

Imagine what it would be like to have to learn the chemical and physical properties of all of the elements and compounds now known. This would be a frightening task, and fortunately we do not have to do it. Chemists have classified the elements in such a way that if we study in detail just one member of a group of similar elements, it is possible to predict the likely behaviour of other elements (and their compounds) in the same group.

One of the important classification systems used in chemistry is the Periodic Table, in which the elements are arranged in order of increasing atomic number, i.e. the number of electrons in their atoms. As we have already seen, the number of electrons in the outer shell of an atom determines the chemical properties of the atom. Each row of elements in the table ends with an element which has atoms containing a full outer shell of electrons. It will be helpful to think of the first part of the Periodic Table (i.e. the first 20 elements) as shown in the Figure. An actual Periodic Table is reproduced on the last page of the book.

Notice that hydrogen is not placed in the same vertical column as other elements; it is unique, and does not really belong to any group.

Vertical columns of elements in the table are called groups.
Horizontal rows of elements in the table are called periods.

Certain parts of the table have special names, as shown below. Make sure that you learn and understand the following.

1 The elements in any group all have the same number of electrons in the outer shells of their atoms (see Figure, p. 90), and they thus have very similar chemical properties. The alkali metals, for example, all have the following properties which make them very similar to each other but very different from most other metals.

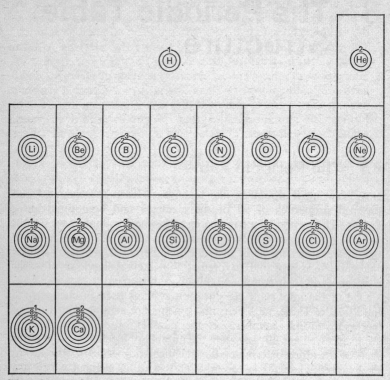

The electronic structure of the first twenty elements in the periodic table

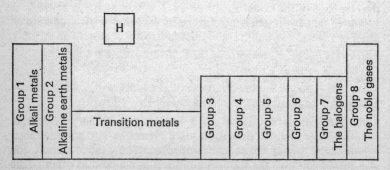

Regions of the Periodic Table

They are less dense than water (float), can be cut with a knife, are kept under oil, melt on a water surface and dart around on it as they react to liberate hydrogen gas and form an alkaline solution.
2 The elements in Group 1 all need to *lose* one electron to form stable electron structures, and consequently they have a combining power (valency) of 1. Similarly elements of Groups 2 and 3 have combining powers of 2 and 3 respectively. Group 4 elements have a valency of 4, although they *share* electrons in order to form stable structures rather than to lose or gain 4 electrons, which is very difficult. The elements in Groups 5, 6 and 7 have valencies of 3, 2 and 1 respectively for they need to *gain* electrons. A consequence of this is that if you know the formula of a compound of an element in, say, Group 3 (e.g. Al_2O_3) then it is possible to state the formula of a similar compound of another Group 3 element, even if you have never heard of it, e.g. Ge_2O_3.
3 The elements on the left hand side of the table tend to be metals, as they have 1, 2 or 3 electrons in their outer shells. Elements on the right hand side tend to be non-metals as they have 5, 6, 7 or 8 electrons in their outer shells. Metallic properties thus increase from left to right across the table.

Changes across a period
Chemical activity
An element of Group 1 is more reactive than the Group 2 element in the same period because it is easier to lose 1 electron (in forming stable electronic structures) than it is to lose 2. This trend continues across the table so that chemical activity is high at the beginning of a period and falls to a low point in Group 4. From this point, elements need to *gain* electrons and as it is easier to gain 2 electrons (Group 6 elements) than 3 electrons (Group 5 elements), reactivity increases again and reaches another peak at Group 7. Group 8 elements have little or no chemical reactivity.

You may have done experiments which show this trend, by comparing the chemical behaviour of each of the elements in a period with the same substance, e.g. water or steam.

The nature of the oxides
The change in type of oxide is a consequence of the increase in metallic behaviour from left to right across a period. As metal oxides are generally basic, oxides of elements on the left hand side of a period tend to be basic. As non-metallic oxides are often acidic or neutral, the oxides of elements on the right hand side of a period

tend to be acidic or neutral. Sometimes this gradual change from left to right across a period produces an 'intermediate' oxide which is amphoteric, e.g. aluminium oxide in Group 3 of Period 3.

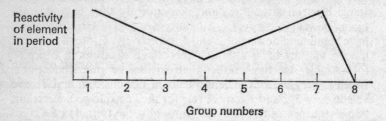

Reactivity trends across a period

Structural changes

Both the elements and their compounds follow the same *general* trend. The metals and their compounds (i.e. those on the left hand side of the table) tend to form giant structures (either metallic or ionic lattices). They usually have high melting points and boiling points, and are solids at room temperature.

On the right hand side the non-metals and their compounds *with other non-metals* are always covalently bonded. Many of these are molecular, in which case they have low boiling and melting points and are usually either liquids or gases at room temperature; if solids, they are easily melted. Some of the substances in this region of the table have giant structures however, even though they are covalent (e.g. diamond, graphite, silicon dioxide), and so they have high melting and boiling points.

The special case of the transition elements

These blocks of elements (e.g. iron, copper manganese, chromium) in the 'centre' of the table are very similar when considered as a *horizontal* row. They (or their compounds) usually have the following properties.

1. The elements are hard, dense metals with high melting and boiling points.
2. The elements often have several valencies, e.g. iron, 2 and 3.
3. Their compounds are often coloured.
4. The elements and their compounds often show catalytic properties.

Changes down a metallic group
Atomic and ionic sizes
As the atomic number increases down a group, the atomic radius of the atoms also increases. An atom of lithium (atomic number 3) is smaller than an atom of sodium (atomic number 11) because it contains fewer shells of electrons.

A metallic ion (i.e. a positive ion) is always smaller than the atom from which it came, because the 'atom' has lost its outer shell of electrons.

The ionic radius also increases down a group of similar elements, because each ion has one more electron shell than the previous one.

Reactivity
All of the members of a group of metals react in a similar way with a certain substance, but do so with increasing vigour as the atomic number increases.

You will probably have seen experiments which show that lithium, sodium and potassium are all very similar in their reaction with water, but that potassium reacts violently whereas lithium reacts more slowly. Similarly, magnesium (Group 2) reacts very slowly with cold water, but calcium reacts more rapidly.

The reason for this is that metals tend to lose electrons and if an outer electron is in a shell relatively far from the nucleus (e.g. the outer electron in an atom of potassium) it has comparatively few empty shells to 'jump through' before it escapes from the atom. The outer electron in an atom of lithium, on the other hand, has more empty shells to jump through before it can escape; this is more difficult and so lithium is less reactive than potassium.

Changes down a non-metal group
Atomic and ionic sizes
Atomic and ionic sizes increase down a group, as with metals. An ion of a non-metal is, however, larger than the atom from which it is formed (c.f. metal ions).

Reactivity
Elements in a group of non-metals again show similar chemical properties to each other, but the vigour of their reactions follows a trend opposite to that shown in metal groups, i.e. they get less reactive as the atomic number increases.

This is because non-metals need to *gain* electrons, and it is comparatively easy to add an electron to a shell which is close to the nucleus because the positive charge on the nucleus is helping to 'pull it in'. For this reason, an element at the top of a group of non-metals is more reactive than those below it.

You may have seen experiments which show this trend, e.g. the reactions of the halogens with iron wool. Other evidence is provided by the displacement reactions of the halogens (p. 161) and by the following data.

$$H_2(g) + F_2(g) \xrightarrow{\text{explosive even in the dark}} 2HF(g)$$
$$H_2(g) + Cl_2(g) \xrightarrow{\text{explosive in sunlight}} 2HCl(g)$$
$$H_2(g) + Br_2(g) \xrightarrow{\text{needs heating}} 2HBr(g)$$
$$H_2(g) + I_2(g) \xrightleftharpoons[\text{then an equilibrium is formed}]{\text{needs heating, and even}} 2HI(g)$$

6.2 Structure

Syllabuses vary considerably with respect to this topic, and it is unlikely that you will need to understand all of the terms and examples which are discussed in this section.

Giant (macro) structures

All of the particles in a giant structure are joined together by strong bonds to form a large, three-dimensional arrangement in which no free, individual units exist. There are three types of giant structures, but as all of the bonds are strong, all types of giant structures are very difficult to 'pull apart' and so have high melting and boiling points, and they are solids at room temperature. The three types can be distinguished, however, by electrical effects. Only metals (and graphite) conduct electricity in the solid state. Only giant ionic structures conduct electricity (and are decomposed by it) when molten or dissolved in water.

1 Metallic structures (metallic lattices)

The atoms of a metal are normally packed in one of two basic ways.

Close packing in metals In close packing, the metal atoms are arranged in flat layers as close together as possible, with the atoms in one layer resting in the hollows in the surface of the layer below. If it is relevant to your syllabus, you should be able to describe the two different ways in which metal particles can be closely packed, i.e. ABAB ... structures and ABCABC ... structures.

Body-centred cubic packing in metals Some metals exist as body-centred cubic structures (see Figure) in which there is more space between the particles than there is in either of the closely packed

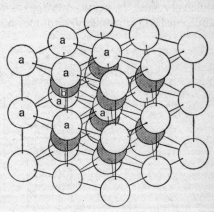

A body centred cubic structure. The spheres are identical but some are shaded for convenience. The spheres marked 'a' form a cube and the sphere b is at the body centre of the cube. This pattern is repeated throughout the structure. For example, the shaded spheres form cubes with unshaded spheres at their body centres

systems. The alkali metals have body-centred cubic structures, and this is partly why they have low densities (their particles are spread out) and are soft.

*The **co-ordination number** of a structure is the number of particles which surround and touch any given particle. If more than one type of particle is present, the co-ordination number for one of them refers to the number of particles of the opposite type which surround and touch it.*

The co-ordination number in closely packed metals is 12, and that in body-centred cubic structures is 8.

2 Giant atomic lattices

These consist of *atoms* bonded together, and so the bonding is covalent. The examples normally encountered are diamond and graphite, the two allotropes of carbon.

***Allotropes** are different forms of the same element in the same physical state.*

Graphite (see Figure) has layers of atoms in hexagonal plates, and all the bonds within a layer are very strong. The bonds between the layers are *relatively* weak, and so the layers can slide over each other. Graphite is thus comparatively soft, feels slippery (it is used suspended in oil as a lubricant) and flakes easily (it is a component of pencil leads). It also conducts electricity.

The atoms in diamond (see Figure) are each strongly bonded by four covalent bonds directed to the corners of a tetrahedron, and there are no flat planes which can slide over each other. Diamond is thus very hard, and is used in rock-cutting drills. It can also be cut and polished (in jewellery) and does not conduct electricity.

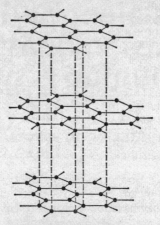

The structure of graphite. The dotted lines represent weak bonds between the planes

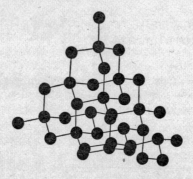

The structure of diamond. Each atom is bonded to those around it by four strong bonds

These contrasting properties of the two allotropes of carbon are neatly explained by their structures, and yet they are both pure forms of carbon. As they are both giant structures they have high melting and boiling points.

3 Giant ionic structures

These are giant structures in which large numbers of ions are held together by opposite charges in a three-dimensional structure. The structure of sodium chloride is shown in the Figure.

The co-ordination number around a sodium ion is 6 and that around a chloride ion is 6; we say that the co-ordination is 6:6. The sodium ions (considered alone) form face-centred cubes and so do the chloride ions. The structure is often described as 'two interpenetrating face-centred cubic lattices, one of sodium ions and one of chloride ions'.

*A **unit cell** is the simplest three-dimensional arrangement which is regularly repeated throughout a structure; the unit cell of sodium chloride is that shown in the Figure below.*

Sodium chloride crystallizes in cubic crystals because this is the arrangement of the ions within it. Similarly, its crystals can only be cleaved (split) along lines which follow the layers of ions, i.e. at right angles to each other.

Note: If the layers of ions are displaced by just one interionic distance, similar charges come opposite one another, so there is *repulsion* between the layers rather than attraction. This is the main reason why ionic substances are so *brittle* when subjected to shock or strain.

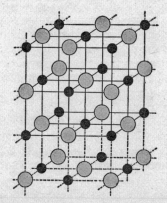

The structure of sodium chloride, NaCl.
The larger spheres represent the chloride ions. The solid lines indicate the unit cell

Molecular structures

It is essential that you understand the points made on page 31 (e.g. why molecular structures have low boiling points) before you read this section; one of the most misunderstood points in elementary chemistry is the difference between *inter*molecular and *intra*molecular bonding. Make sure also that you contrast molecular structures with other covalent structures such as the giant atomic lattice of graphite.

Common examples of molecular structures are ice, iodine and sulphur. In all these, the *intra*molecular bonds, those holding the atoms together in the molecule, are strong and cannot easily be broken. In ice, the molecules of water are held together to form a structure which takes up more volume than the liquid water which produced it; water expands on freezing, unlike most other liquids. The *inter*molecular bonds are easily broken when ice is heated, and the molecules then separate, i.e. it melts. In iodine, the *inter*molecular forces between I_2 molecules are also easily broken by heating, and in this case the substance sublimes as the molecules come apart from each other.

Solid sulphur is often studied in more detail, for although its molecules are always composed of 8 atoms (S_8) arranged in a 'puckered ring', it is possible to pack the molecules in two different ways. Solid sulphur thus has two allotropes, the α (rhombic) and β (monoclinic) forms. You could be asked to describe experiments to prepare samples of these allotropes. The way in which the molecules pack together in the two allotropes (c.f. the way the atoms of carbon are arranged in diamond and graphite) is unimportant, but you will probably have seen or drawn the different appearances of their crystals and realize that they have different densities.

You may also have seen an experiment in which 'plastic' sulphur is made. This is not really a true allotrope of sulphur, and it changes fairly quickly into one of the other forms. The experiment is of interest, however, for it shows that sulphur undergoes several structural changes when heated. When the solid is heated *gently*, it first melts to a mobile pale yellow liquid. On further heating, the colour darkens and the viscosity increases until the liquid becomes very dark and very viscous. Suddenly, near the boiling point, the liquid becomes mobile again. These observations are relevant to some syllabuses, but you should not attempt to *explain* them unless you have discussed the changes (as structural changes) in your course.

The Periodic Table. Structure 99

Questions

1 In any one group of the Periodic Table, the elements: (*a*) become more reactive as their atomic numbers increase; (*b*) become less reactive as their atomic numbers increase; (*c*) lose electrons more easily as their atomic numbers increase; (*d*) lose electrons less readily as their atomic numbers increase; (*e*) are all equally reactive.

2 Element X is in Group 1 of the Periodic Table. It is likely to be: (*a*) a very reactive non-metal; (*b*) an element which readily forms X^- ions; (*c*) a dense, hard metal with a high melting point; (*d*) a light, soft metal with a low melting point; (*e*) a dense, soft metal with a high melting point.

3 The element with atomic number 10 is likely to have similar properties to the element with atomic number: (*a*) 9; (*b*) 11; (*c*) 16; (*d*) 18; (*e*) 28.

4 The following are ionic and atomic radii (in nm) of members of the same group of the Periodic Table.

	Atomic radius	Ionic radius
A	0.133	0.078
B	0.157	0.098
C	0.203	0.133
D	0.216	0.149
E	0.235	0.165

(*a*) Is this a metallic group or a non-metallic group?
(*b*) Which element would have the lowest atomic number?
(*c*) Which element would be the most reactive?

5 (*a*) Sodium and aluminium have atomic numbers of 11 and 13 respectively. They are separated by one element in the Periodic Table, and have valencies of one and three respectively. Chlorine and potassium are also separated by one element in the Periodic Table (they have atomic numbers of 17 and 19 respectively) and yet they both have a valency of one. Explain the difference.

(*b*) The halogens, fluorine, chlorine, bromine and iodine, show a gradation in properties. Illustrate this by reference to their ease of combination with hydrogen, and the ease of replacement of one halogen by another. (JMB)

6 A solid melts at 1020 °C and is an electrolyte when molten. It has: (*a*) a molecular structure; (*b*) a giant atomic structure; (*c*) a metallic structure; (*d*) a giant ionic structure; (*e*) some other structure.

7 Which of the following pairs are allotropes?
(*a*) Carbon (graphite) and sulphur (monoclinic).
(*b*) Carbon (diamond) and sulphur (monoclinic).
(*c*) Sodium chloride and caesium chloride.
(*d*) Phosphorus (white) and phosphorus (red).
(*e*) ^{31}P and ^{30}P.

8 The data in the table refer to five *elements* lettered V, W, X, Y, Z:
(*a*) Which pair of elements are metals?
(*b*) Which pair are allotropic forms of the same element?
(*c*) Which element could be an inert gas?

Element	Atomic mass	Melting point K (°C)	Electrical conductivity of solid element	Reaction with oxygen
V	31	863 (590)	Non-conductor	Burns readily
W	40	1124 (851)	Good	Burns readily
X	20	24 (−249)	Non-conductor	No reaction
Y	207	600 (327)	Good	Oxidizes slowly
Z	31	317 (44)	Non-conductor	Burns readily

(WEL)

7 The mole. Chemical formulae and equations

7.1 The mole

If different elements are weighed out in the same proportions as their atomic masses (e.g. 1 g of hydrogen, 12 g of carbon) it can be shown that the quantities so obtained contain the same number of *atoms*. This number is known to be 6.02×10^{23}, and is called the **Avogadro number**. Similarly, the **molecular mass** of any compound (expressed in grams) contains 6.02×10^{23} *molecules* of the substance, and the 'formula mass' of any ionic substance (expressed in grams) contains 6.02×10^{23} *formula units* of the substance.

In the same way that we buy eggs in packs of 6, chemists count particles in packs of 6.02×10^{23}. Each unit pack of particles (i.e. 6.02×10^{23}) is called a mole. The unit pack has to contain a large number of particles or otherwise we would not be able to work with them conveniently, e.g. by weighing.

The amount of substance containing 6.02×10^{23} particles is a basic scientific unit called the mole.

The mole is an amount of substance usually expressed as a mass in grams. As 12 g of carbon contains 6.02×10^{23} atoms of carbon, it follows that a mole of carbon atoms (i.e. a unit pack of carbon atoms) has a mass of 12 g. The term can be applied in exactly the same way to compounds, taking care that we refer to the correct type of particle in each case, e.g. molecules or formula units. For example: a mole of water (H_2O) has a mass of 18 g $(1 + 1 + 16)$ and contains 6.02×10^{23} *molecules* of water; a mole of sodium hydroxide (NaOH) has a mass of 40 g $(23 + 16 + 1)$ and contains 6.02×10^{23} *formula units* of sodium hydroxide.

Note:
1 The statement '16 g of oxygen' could mean 16 g of oxygen atoms, 16 g of oxygen molecules (O_2) or 16 g of oxygen ions (O^{2-}).

Always state the type of particle being considered and make sure that you understand the difference between I_2 (an iodine molecule), $2I$ (two atoms of iodine) and $2I^-$ (two iodide ions). In the absence of any clear statement about the type of particle being considered, you must assume that the substance is in the form in which you would normally find it. Thus '16 g of oxygen' would be taken to mean 16 g of oxygen molecules (O_2), as we normally encounter oxygen as O_2 molecules.

2 The term 'formula unit' can be applied to all compounds, whether they are made up of molecules, ions or have more complicated structures. However it is only correct to refer to a mole of molecules if the compound is actually made up of molecules.

3 The term 'mole of ions' can also be used, but it refers to 6.02×10^{23} ions *of the same type*, e.g. a mole of sodium ions Na^+. It cannot, therefore, be used to describe a *solid* ionic compound, which must be made up of at least two kinds of ion. The term is useful when considering solutions of ionic compounds, in which ions separate from each other. In the solid ionic compound sodium chloride we refer to a mole of sodium chloride units (NaCl), and in solution these ions separate from each other to produce a mole of sodium ions *and* a mole of chloride ions.

How a chemist uses moles

A chemist uses moles as 'measurable packages of particles' in order to understand how *individual* particles take part in chemical reactions. Thus if it can be shown that 12 g of carbon react exactly with 32 g of oxygen (both of which are readily measurable quantities) we can say

  1 mole (12 g) of carbon atoms react with

  $\qquad\qquad\qquad\qquad$ 1 mole (32 g) of oxygen molecules.

  6.02×10^{23} atoms of C react with 6.02×10^{23} molecules of O_2.

This is the same ratio as 1 atom of carbon reacts with 1 molecule of oxygen, and we can write

$$C(s) + O_2(g) \to CO_2(g)$$

We have constructed a chemical equation to show how *individual* atoms and molecules react (even though we cannot see them or weigh them) by working in moles, which are manageable and measurable quantities.

Molar solutions

A molar solution contains one mole of dissolved substance per dm^3 of solution.

For example, a molar solution of substance A contains 6.02×10^{23} particles of A per dm^3 of solution. Thus when 58.5 g of sodium chloride (23 + 35.5) is dissolved in water and made up to 1 dm^3 of solution, a molar solution of the compound is produced and is recorded as having a concentration of 0.1 mol dm^{-3} (or 0.1 M).

Just as it is important to state the type of particle being referred to when using moles, it is equally important to state the substance being referred to when using molarities. Thus 106 g of anhydrous sodium carbonate, Na_2CO_3 [(23 + 23 + 12 + (3 × 16))] dissolved in 1 dm^3 of solution can be described as having a concentration of 1 mol dm^{-3} (or 1 M) with respect to sodium carbonate. It is equally correct to say that the solution has a concentration of 1 mol dm^{-3} (1 M) with respect to carbonate ions (i.e. it contains one mole of carbonate ions per dm^3 of solution), and it also has a concentration of 2 mol dm^{-3} (2 M) with respect to sodium ions. This is because each dissolved Na_2CO_3 unit produces 2 sodium ions and 1 carbonate ion.

Basic calculations with molarities

An understanding of the following types of manipulation is needed for volumetric analysis calculations, which are considered in Unit 11.

(a) What mass of sodium hydroxide must be dissolved in 250 cm^3 of solution to produce a solution which has a concentration of 0.1 mol dm^{-3} (0.1 M) with respect to sodium hydroxide?

Answer 1 dm^3 of 0.1 M NaOH solution requires 0.1 moles NaOH.
  1 mole NaOH is 40 g (23 + 16 + 1).
  1 dm^3 0.1 M solution thus requires 0.1 × 40 g.
  ∴ 250 cm^3 ($\frac{1}{4}$ dm^3) 0.1 M solution needs
$$\tfrac{1}{4} \times 0.1 \times 40 \text{ g} = 1 \text{ g}.$$

(b) How many moles of formula units of hydrochloric acid are contained in 25 cm^3 of a solution which has a concentration of 0.1 mol dm^{-3} (0.1 M) with respect to the acid?

Answer 1 dm^3 of a 0.1 M solution of the acid contains 0.1 moles of formula units.
  ∴ 1 cm^3 of 0.1 M acid solution contains
$$\frac{0.1}{1000} \text{ moles of formula units.}$$
  ∴ 25 cm^3 of 0.1 M acid solution contains
$$\frac{25 \times 0.1}{1000} \text{ moles formula units} = 0.0025 \text{ moles formula units.}$$

(c) How many moles of chloride ions are contained in 250 cm³ of a solution which has a concentration of 0.5 mol dm⁻³ (0.5 M) with respect to magnesium chloride, $MgCl_2$?

Answer 1 mole of $MgCl_2$ contains 2 moles of Cl^-(aq) ions.
∴ 1 dm³ of a 0.5 M solution contains
$$2 \times 0.5 \text{ moles } Cl^-(aq).$$
∴ 250 cm³ ($\frac{1}{4}$ dm³) of 0.5 M solution contains
$$\tfrac{1}{4} \times 2 \times 0.5 \text{ moles } Cl^-(aq) = 0.25 \text{ moles of } Cl^-(aq).$$

(d) What is the concentration (in mol dm⁻³) of a solution containing 2 g of sodium hydroxide in 500 cm³ solution?

Answer 2 g in 500 cm³ = 4 g per dm³.
The molar mass of sodium hydroxide is 40.
4 g is 0.1 moles.
∴ 0.1 moles are dissolved per dm³, and the solution has a concentration of 0.1 mol dm⁻³ (0.1 M).

7.2 Percentage composition. Formulae

The laws of chemical combination

*The **law of conservation of mass** states that matter is neither created nor destroyed in a chemical reaction.*

Some syllabuses could require details of an experiment to illustrate this law.

*The **law of constant composition** (or **definite proportions**) states that all pure samples of the same chemical compound contain the same elements combined together in the same proportions by mass.*

Thus all pure samples of copper(II) oxide (no matter how made, are found to contain (by mass) 80% of copper and 20% of oxygen. Calculations of this type are discussed under the next heading.

*The **law of multiple proportions** states that when two elements A and B combine together to form more than one compound, the different masses of A which combine with a fixed mass of B are in a simple ratio.*

Some syllabuses could require you to verify this law by using experimental data provided in a question. If faced with a problem of this type, first use the data to produce statements of the following type (one statement for each compound).

12 g of element A combine with 10 g of element B in compound 1.
9 g of element A combine with 15 g of element B in compound 2.

Then rewrite the statements in such a way that the mass of B in each statement is fixed. The above statements become

36 g of A combine with 30 g of B in compound 1.
18 g of A combine with 30 g of B in compound 2.

The different masses of A which combine with this fixed mass of B should be in a simple ratio, in this case 36:18 or 2:1.

Percentage composition of compounds

In order to calculate the percentage composition by mass of a compound, proceed as follows (using ammonium nitrate as an example).

1 Write down the formula of the compound: NH_4NO_3.
2 Find the formula mass: $14 + (4 \times 1) + 14 + (3 \times 16) = 80$.
3 Express each atomic mass as a percentage of the formula mass. (If more than one atom of a particular element is present, then the total mass of these atoms must be used, e.g. as two nitrogen atoms appear in the formula, the mass used is 28 and not 14.)

$$\text{per cent nitrogen} = \frac{28}{80} \times 100\% = 35\%$$

$$\text{per cent hydrogen} = \frac{4}{80} \times 100\% = 5\%$$

$$\text{per cent oxygen} = \frac{48}{80} \times 100\% = 60\%$$

4 Check that the percentages add up to 100.

Note: If a hydrated compound is being considered, the percentage of water should be calculated as a complete separate unit rather than considering the water as the individual elements hydrogen and oxygen. Thus in hydrated magnesium chloride, $MgCl_2.6H_2O$, formula mass 203, the percentage composition by mass is

$$\text{per cent magnesium} = \frac{24}{203} \times 100\% = 11.82\%$$

$$\text{per cent chlorine} = \frac{71}{203} \times 100\% = 34.98\%$$

$$\text{per cent water} = \frac{108}{203} \times 100\% = 53.2\%$$

Experiments to determine the percentage of water of crystallization

You will probably have done an experiment of this type. The principle is to weigh a suitable container empty, and then again when containing some of the compound, so as to determine the mass of the compound. This is then heated gently until no further change in mass occurs, when the loss in weight is equal to the mass of the water of crystallization given off. Then, percentage of water of crystallization (by mass)

$$= \frac{\text{mass of water of crystallization}}{\text{mass of hydrated compound}} \times 100\%$$

Formulae

You should be able to work out the formula of any common compound by using your knowledge of symbols and valencies. However, it is also important to be able to work out the formula of a compound from experimental data, which is what happens when a new compound is discovered. Chemists use moles to calculate a formula from experimental results. The experimental results could be given as the masses of the different chemicals which combine with each other, or as percentages (by mass) of the different chemicals in the compound.

Empirical formulae

In order to determine the empirical (simplest ratio) formula of a compound from experimental data, proceed as follows.

1 Use the data to write a statement about the masses of the elements which combine together, e.g.
8.32 g of lead combine with 1.28 g sulphur and 2.56 g oxygen.
(Percentage compositions are used in the same way. Thus if a compound contains 40% Cu, 20% S and 40% O, we would write: 40 g of copper combine with 20 g of sulphur and 40 g of oxygen.)
Note: The data which will enable you to write a statement like this may be given in an indirect way, and will first have to be 'unravelled'. Examples of this are given after the basic procedure so as not to confuse the calculation.

2 Rewrite the statement by dividing the mass of each element by the mass of one mole of atoms of that element, so that we are now working in moles of atoms.

$\frac{8.32}{207}$ *moles of lead atoms combine with* $\frac{1.28}{32}$ *moles of sulphur atoms and* $\frac{2.56}{16}$ *moles of oxygen atoms.*

∴ *0.04 moles Pb atoms combine with 0.04 moles S atoms and 0.16 moles O atoms.*

3 Convert these mole ratios into whole numbers by dividing each of them by the smallest (0.04 in this case).

∴ *1 mole of Pb atoms combines with 1 mole S atoms and 4 moles O atoms.*
∴ *1 atom of Pb combines with 1 atom of S and 4 atoms of O and the empirical formula is $PbSO_4$.*

It is important to set out all calculations clearly, using a new line for each step. It is possible to gain good marks in a chemical calculation even if your final answer is incorrect, *providing that you show your working clearly and show that you understand the chemistry*. In the above example, the statements you would be expected to write down in your answer are printed in italics.

Note that the formula obtained in this way is not necessarily the actual formula (i.e. the molecular formula); it is only the ratio in which the atoms combine. You must always make this clear by referring to it as the empirical formula.

Sometimes the data needed to produce the first statement in the argument are 'disguised' as experimental readings. These usually take the form of weighings obtained by combining chemicals together (e.g. by burning magnesium in air to make magnesium oxide) or by breaking down a chemical (e.g. by reducing a metal oxide to the metal in a stream of hydrogen). If you have used an experiment of this type you should be able to describe in detail how you obtained an empirical formula from it. The essential thing is to use the weighings from the experiment to determine the masses of the chemicals which actually *react* together, and then to proceed as shown previously.

Molecular formulae

In order to proceed from an empirical formula to the actual (molecular) formula, the molecular mass of the compound is needed. This may be given directly as a statement of fact, or indirectly, e.g. in the form of a vapour density (p. 215), from which 2 × vapour density = molecular mass.

Suppose that an empirical formula of HO has been determined as in the previous section, and the molecular mass is 34. The molecular formula is always a simple multiple of the empirical formula, or the same as the empirical formula.

If the empirical formula = HO, the molecular formula = $(HO)_x$, = H_xO_x (where x is a small whole number).
∴ *Molecular mass of $H_xO_x = x + 16x = 17x = 34$.*
∴ *$x = 2$ and the molecular formula is H_2O_2.*

Once again, the statements you would be expected to write down in your answer are shown in italics. Remember that sometimes the empirical formula is the same as the molecular formula, but you always need the molecular mass in order to find out.

7.3 Equations

Constructing and balancing equations

Experimental evidence has enabled us to write an equation for every chemical reaction. You will be expected to either learn all of the equations you have encountered in your own course, or to be able to work them out from basic principles. Many students have difficulty with equations, but it must be remembered that a fully balanced equation is a very important piece of chemistry. As a general rule, every chemical reaction you describe in an examination answer should be given an equation, and there will often be a mark allowed for each equation that is relevant.

Ionic equations

Consider the equations which show the results of a chloride test on separate solutions of sodium chloride and potassium chloride.

$$NaCl(aq) + AgNO_3(aq) \rightarrow AgCl(s) + NaNO_3(aq)$$
$$KCl(aq) + AgNO_3(aq) \rightarrow AgCl(s) + KNO_3(aq)$$

These are perfectly correct, properly balanced equations, but they are also slightly misleading as they suggest that sodium nitrate and potassium nitrate have been *formed*. In fact they have not, for the solution formed contains separate sodium and nitrate (or potassium and nitrate) ions; these ions have not combined. The starting solutions also contained these separate ions; they have not changed during the course of the reaction and it would be better not to include them in the equation. (They are known as *spectator ions*.)

Silver chloride has been *formed*, however, during the course of the reaction, for although the initial solutions contained separate silver and chloride ions, these join together to form a precipitate of silver chloride in the reaction. This change should be shown in the equation. A better equation for the reaction is therefore the ionic one:

$$Ag^+(aq) + Cl^-(aq) \rightarrow AgCl(s)$$

Ionic equations are extremely useful in chemistry, for they are shorter than full equations, give a more accurate idea of what has actually changed in a reaction, and often one ionic equation can serve to summarize many apparently different reactions (e.g. the example above is the ionic equation for any positive chloride test in solution). Ionic equations cannot be applied to every reaction but will be applicable to any reaction which involves a solution of an ionic compound as one of the reactants. Examples have been given wherever possible in this book.

Remember that ionic equations must be balanced, and that this includes negative or positive charges. Thus

$$Fe(s) + Ag^+(aq) \to Fe^{2+}(aq) + Ag(s)$$

is not balanced, but

$$Fe(s) + 2Ag^+(aq) \to Fe^{2+}(aq) + 2Ag(s)$$

is balanced.

The interpretation of equations

A properly balanced equation provides a great deal of information. It shows the molar ratios in which the chemicals react, and from these we can determine the masses that would react together in a certain situation, and calculate the mass (or volume, if a gas) of product that would be formed. The state symbols also tell us about any changes of state which might occur during a reaction, and the ΔH sign (if present) tells us whether the reaction is endothermic or exothermic. The $\rightleftharpoons$ sign, if present, also provides valuable information.

Determining reacting masses from equations

Suppose that we are asked to calculate the mass of magnesium oxide produced by the complete combustion of 9.6 g of magnesium in pure oxygen. (The actual statements that you would be expected to show in an examination answer are shown in italics.)

1 Write down a fully balanced equation for the reaction. (This is vital; you cannot perform these calculations without an equation.)

   $2Mg(s) + O_2(g) \to 2MgO(s)$

2 From the equation, write down the numbers of moles involved with each of the substances mentioned in the problem.

   2 moles of magnesium atoms $\to$ 2 moles magnesium oxide.

3 Convert moles into masses.

   48 g (2 × 24) magnesium $\to$ 80 g (2 × 40) magnesium oxide

   and then proceed as follows,

4 1 g magnesium $\rightarrow \dfrac{1}{48} \times 80$ g magnesium oxide.

9.6 g magnesium $\rightarrow 9.6 \times \dfrac{1}{48} \times 80$ g magnesium oxide.

9.6 g magnesium $\rightarrow 16$ g magnesium oxide.

There are a few variations of this type of question, but the basic idea is the same. Suppose that we are asked to calculate the mass of calcium hydroxide that will react with 21.4 g of ammonium chloride, and also the mass of the ammonia produced. Using the steps as in the previous example,

$$2NH_4Cl(s) + Ca(OH)_2(s) \rightarrow 2NH_3(g) + CaCl_2(s) + 2H_2O(g)$$

2 moles ammonium chloride react with 1 mole calcium hydroxide $\rightarrow$ 2 moles ammonia.

107 g NH_4Cl react with 74 g $Ca(OH)_2 \rightarrow 34$ g NH_3.

1 g NH_4Cl react with $\dfrac{1}{107} \times 74$ g $Ca(OH)_2 \rightarrow \dfrac{1}{107} \times 34$ g NH_3.

21.4 g NH_4Cl react with $21.4 \times \dfrac{1}{107} \times 74$ g $Ca(OH)_2 \rightarrow$

$$21.4 \times \dfrac{1}{107} \times 34 \text{ g } NH_3.$$

21.4 g NH_4Cl react with 14.8 g $Ca(OH)_2$ to produce 6.8 g NH_3.

Determining reacting volumes from equations

These calculations, although similar to those involving masses, may also require the conversion of the volume of a gas under a certain set of conditions to the volume it would occupy under different conditions, and so are considered in Unit 11.

Reactions between solids and standard solutions

The procedure is as before except that this time concentrations are involved. Suppose that we are asked to calculate the concentration of a hydrochloric acid solution, 100 cm³ of which dissolves 3.0 g of magnesium ribbon.

$$Mg(s) + 2HCl(aq) \rightarrow MgCl_2(aq) + H_2(g)$$

1 mole magnesium atoms react with 2 moles hydrochloric acid.

∴ 24 g Mg react with 2 moles HCl.

∴ 1 g Mg react with $\dfrac{1}{24} \times 2$ moles HCl.

∴ 3 g Mg react with $3 \times \dfrac{1}{24} \times 2$ moles = 0.25 moles HCl.

∴ 0.25 moles HCl must be present in 100 cm³ solution.
∴ 2.5 moles HCl must be present in 1000 cm³ solution.
The acid has a concentration of 2.5 mol dm⁻³ (2.5 M).

Suppose that we are asked to calculate the mass of pure iron that would be dissolved by 500 cm³ of 0.1 M sulphuric acid.

$$Fe(s) + H_2SO_4(aq) \rightarrow FeSO_4(aq) + H_2(g)$$

1 mole iron atoms react with 1 mole sulphuric acid.
∴ 56 g of iron atoms react with 1 mole of sulphuric acid.
But 500 cm³ of 0.1 M sulphuric acid contain

$$\tfrac{1}{2} \times 0.1 \text{ moles} = 0.05 \text{ moles}.$$

∴ 56 × 0.05 g of iron react with 0.05 moles of the acid = 2.8 g.

Using precipitation reactions to determine equations

You may have seen how a precipitation reaction can be used to determine a chemical equation. The following example illustrates the method and will also help to revise some important ideas.

Solutions of a metal sulphate and of barium chloride ($BaCl_2$), each of concentration 1 mol dm⁻³ (1 M), were used in the reaction. Five tubes were taken, containing the following mixtures.

Tube	Volume of metal sulphate solution (cm³)	Volume of barium chloride solution (cm³)	Volume of water (cm³)	Total volume of mixture (cm³)
1	2	2	8	12
2	2	4	6	12
3	2	6	4	12
4	2	8	2	12
5	2	10	0	12

A precipitate (of barium sulphate, $BaSO_4$) appeared in each tube. The tubes were centrifuged and the heights of the precipitate were recorded as follows.

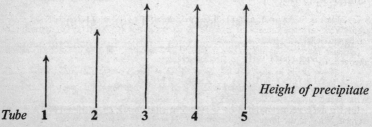

As the precipitate remained constant in tubes 3, 4 and 5, we can conclude that the chemicals mixed in tube 3 were sufficient to complete the precipitation, and that further additions merely provided an excess of barium ions. In tube 3, 2 cm³ M sulphate solution reacted with 6 cm³ M $BaCl_2$ solution.

∴ 2 dm³ M sulphate solution would react with 6 dm³ M $BaCl_2$ solution.
∴ 2 moles metal sulphate react with 6 moles $BaCl_2$.
∴ 2 moles of metal sulphate react with 6 moles Ba^{2+} ions.
Each mole Ba^{2+} ions react with 1 mole SO_4^{2-} ions

$$(Ba^{2+}(aq) + SO_4^{2-}(aq) \rightarrow BaSO_4(s)).$$

∴ 2 moles metal sulphate must contain 6 moles SO_4^{2-} ions.
1 mole metal sulphate must contain 3 moles SO_4^{2-} ions.

As a sulphate ion has a combining power of 2, the metal ion must have a combining power of 3 and the formula of the sulphate is $X_2(SO_4)_3$.

The experiment shows us that the metal sulphate and the barium chloride solution react in the ratio 1 mole:3 moles, so we can write

$$X_2(SO_4)_3(aq) + 3BaCl_2(aq) \rightarrow$$

Balancing the equation in the usual way, this becomes

$$X_2(SO_4)_3(aq) + 3BaCl_2(aq) \rightarrow 2XCl_3(aq) + 3BaSO_4(s)$$

Questions

1 One mole of carbon dioxide: (*a*) contains the Avogadro Number of carbon dioxide atoms; (*b*) contains the same number of molecules as 1 g of hydrogen; (*c*) contains the same number of molecules as 16 g of oxygen; (*d*) has a mass of 28 g; (*e*) contains 6.02×10^{23} molecules of carbon dioxide (C = 12; O = 16).

2 25 g of nitrogen: (*a*) contains 6×10^{23} atoms of nitrogen; (*b*) contains 6×10^{23} molecules of nitrogen; (*c*) contains two moles of nitrogen; (*d*) is heavier than 28 g of hydrogen; (*e*) contains the same number of atoms of nitrogen as 35.5 g of chlorine (N = 14; Cl = 35.5).

3 (*a*) What is the mass of a mole of (i) calcium nitrate; (ii) zinc sulphate?
(*b*) What are the masses of a mole of (i) propanol (C_3H_7OH); (ii) tetrachloromethane (CCl_4); (iii) benzene (C_6H_6)?
(*c*) How many moles are present in (i) 100 g of calcium; (ii) 13 g of zinc; (iii) 28 g of nitrogen; (iv) 8 g of S?
(C = 12; H = 1; Ca = 40; N = 14; O = 16; S = 32; Zn = 65; Cl = 35.5).

4 (*a*) 3.5 g of nitrogen combined with 2 g of oxygen to form an oxide of nitrogen. What is the formula of this oxide?
(*b*) On analysis, a compound was found to contain 2.3 g of sodium and 0.8 g of oxygen. What is the formula of the oxide? (N = 14; O = 16; Na = 23).

5 Calculate the empirical formula of a compound that has the composition: 52.0% zinc; 9.6% carbon; 38.4% oxygen. (AEB)

6 An organic compound was found to contain 12.8% carbon, 2.1% hydrogen and 85.1% bromine. The vapour density of the compound was 94.
(a) What is the empirical formula of the compound?
(b) What is the molecular mass of the compound?
(c) What is the molecular formula of the compound? (JMB)

7 A gas has a vapour density of 13.5, and contains 3.70% hydrogen and 44.44% carbon by mass. Analysis shows the gas to contain hydrogen, carbon and nitrogen only. Calculate the molecular formula of this gas. (JMB)

8 When 3.1 g of a carbonate MCO_3 are heated to constant mass, 2.0 g of the oxide of the metal are formed. On reduction of this oxide, 1.6 g of pure metal remains.
(a) Describe experiments by which these figures could be determined in the laboratory.
(b) Give equations for the two reactions.
(c) Calculate the atomic mass of M.
(d) Calculate the volume of carbon dioxide evolved during the heating of the carbonate. (AEB)

8 The metals and their compounds

There are more facts to learn in this Unit than in most of the others. It is important, therefore, to make the task as easy as possible for revision purposes. When you need to remember something about a particular metallic compound (e.g. calcium carbonate) try to think in the following way.

1 Ask yourself what *type* of compound it is (in this case a metallic carbonate) and then try to remember the *general* properties of this type and whether the named example is exceptional in any way. It is far easier to do this than to learn the properties of every metallic compound you have studied; in a typical course this might involve the study of 70 or more metallic compounds.

2 Then look at it from 'the other way round', e.g. calcium carbonate is also a compound of *calcium* as well as a carbonate, and if you remember the general properties of calcium compounds you should be able to add further to your knowledge of this particular substance. For example, as all calcium compounds are white and ionic, then so is calcium carbonate.

If you do not think of *types* of chemicals in this way, there is a real danger that you could fail to answer a question which is in fact quite easy. Suppose that you are asked to describe the preparation of a pure sample of zinc sulphate. Many students would rack their brains trying to think of how they have prepared the substance in the laboratory, and this may be the wrong approach for it is quite possible that they may never have prepared this particular chemical. They are making the mistake of thinking of it as a specific chemical rather than as a general type. If we ask ourselves what *type* of chemical we are referring to, it becomes a problem of how to prepare a metallic sulphate, i.e. as we learn from the general properties of sulphates, a salt. Our general knowledge of sulphates tells us that zinc sulphate is a soluble salt, and so we can pick any one of several methods of preparation, such as sulphuric acid + zinc oxide (acid +

base), sulphuric acid + zinc hydroxide (acid + base) or sulphuric acid + zinc carbonate (acid + carbonate). The *details* of the method we choose will be the same as when we used the method in the laboratory, but perhaps with different chemicals.

The work in this Unit has been divided into sections where the emphasis is on the non-metal part of a compound (e.g. sulphates, chlorides) and other, smaller, sections where the emphasis is on the metallic part of the compound (e.g. properties of calcium compounds). If you learn these facts conscientiously, application of 'both points of view' (as described above) should enable you to remember the essential chemistry of all the common metallic compounds without learning each one individually. The tables in this Unit list metals in the same order as that of the activity series (except Table 8.3) because the properties of their compounds often follow a similar pattern of reactivity changes.

Some uses of the metals and their compounds are also given. Do not learn all of these; you will recognize those which have been stressed on your own course. It is unlikely that you will be asked to write, for example, a list of the uses of calcium compounds, but more likely that you could be asked to state one important use of a particular metal or its compounds.

8.1 Metallic oxides

Methods of preparation

(a) Direct combination with oxygen This is achieved by heating the metal in the air or oxygen. This is not a good nor convenient method, for some metals only form a thin layer of oxide on the metal (see individual details in table 3.4, page 58).

(b) Heating a metal carbonate This is a useful method, particularly for oxides of less reactive metals, the carbonates of which decompose easily, e.g.

$$CuCO_3(s) \rightarrow CuO(s) + CO_2(g)$$

Carbonates of more reactive metals either do not decompose when heated (Na, K) or decompose only when heated to a high temperature (Ca). The carbonates of aluminium and iron(III) do not exist.

(c) Heating a metal nitrate This is a useful method, particularly for the oxides of metals below calcium in the activity series, the nitrates of which readily decompose, e.g.

$$2Pb(NO_3)_2(s) \rightarrow 2PbO(s) + 4NO_2(g) + O_2(g)$$

As nitrogen dioxide is toxic, these reactions should be carried out in a fume cupboard.

Nitrates of elements at the two 'extremes' of the activity series behave differently when heated. The nitrates of sodium and potassium produce the nitrite and oxygen, e.g.

$$2NaNO_3(s) \rightarrow 2NaNO_2(s) + O_2(g)$$

and the nitrates of silver and mercury produce the metal.

(d) Heating a metal hydroxide This is not as useful as (b) and (c) because the hydroxides are not as readily available, e.g.

$$Cu(OH)_2(s) \rightarrow CuO(s) + H_2O(g)$$

Again, this is suitable for the oxides of less reactive metals; the hydroxides of sodium and potassium do not decompose on heating, and that of calcium requires a very high temperature.

A summary of oxide preparations is given in Table 8.1.

Table 8.1 *A summary of oxide preparations*

	Reaction with oxygen	Heat on carbonate	Heat on nitrate	Heat on hydroxide
K	↑ Not Recommended ↓	×	×	×
Na		×	×	×
Ca		✓	✓	✓
Mg		✓	✓	✓
Al		Carbonate does not exist	✓	✓
Zn		✓	✓	✓
Fe(II)		✓	Rarely used	✓
Fe(III)		Carbonate does not exist	✓	✓
Pb		✓	✓	✓
Cu		✓	✓	✓
Hg		—	×	—
Ag		—	×	—

Properties of metallic oxides

1 The only important general property of metallic oxides is their ability to act as a base. A few (e.g. Al_2O_3, ZnO) can also act as an acid and are amphoteric.
2 The following specific details are sometimes needed for examination questions.
   (a) Mercury(II) oxide behaves in an unusual way when heated, forming mercury and oxygen:

$$2HgO(s) \rightarrow 2Hg(l) + O_2(g)$$

(b) Calcium oxide reacts violently with water. If water is added to a small sample of the oxide (often formed by roasting a piece of limestone in air), an exothermic reaction occurs, steam is formed, and a white powder is formed which takes up a greater volume than the original solid. This is the 'slaking of lime' and the product is calcium hydroxide,

$$CaO(s) + H_2O(l) \rightarrow Ca(OH)_2(s)$$

3 The formulae, etc. of the common metallic oxides are given in Table 8.2.

Table 8.2 *The common metallic oxides*

Formula	Name	Colour	Revision points to note
Na_2O	sodium monoxide	white	Reacts with water to form sodium hydroxide
Na_2O_2	sodium peroxide	white	Normally only encountered as a typical peroxide
CaO	calcium oxide	white	Reacts with water (as above). Good drying agent for ammonia
MgO	magnesium oxide	white	— — — — — — — — — — —
ZnO	zinc oxide	white	Yellow when hot. Amphoteric
Al_2O_3	aluminium oxide	white	Amphoteric
FeO	iron(II) oxide	black	— — — — — — — — — — —
Fe_2O_3	iron(III) oxide	red	— — — — — — — — — — —
Fe_3O_4	iron(II) diiron(III) oxide (tri-iron tetroxide)	black	Magnetic
PbO	lead(II) oxide	yellow	Amphoteric, but usually only encountered as a typical base. The most stable oxide of lead
PbO_2	lead(IV) oxide	brown	An oxidizing agent, like MnO_2; e.g. oxidizes concentrated hydrochloric acid to chlorine
Pb_3O_4	dilead(II) lead(IV) oxide (tri-lead-tetroxide)	red	Behaves as if a mixture of $2PbO + PbO_2$; when dilute nitric acid is added the PbO reacts and dissolves to form soluble lead(II) nitrate, leaving the insoluble PbO_2. Also an oxidizing agent as it appears to contain PbO_2. When heated strongly forms PbO + oxygen
Cu_2O	copper(I) oxide	red	Normally only encountered as the red precipitate in a positive Fehling's test for a reducing sugar
CuO	copper(II) oxide	black	— — — — — — — — — — —
HgO	mercury(II) oxide	red	Decomposes when heated (p. 115)

8.2 Metallic hydroxides

Methods of preparation

(a) The action of water on a metal oxide This is not a general method, and it is restricted to the oxides of metals near the top of the activity series, i.e. potassium, sodium and calcium, e.g.

$$CaO(s) + H_2O(l) \rightarrow Ca(OH)_2(s)$$

(b) Precipitation This is an excellent method for preparing metal hydroxides except for those of sodium and potassium, which are soluble. Mix together two solutions, one containing OH^- ions (i.e. sodium hydroxide solution or ammonia solution) and the other con-

Table 8.3 *The common metallic hydroxides*

Formula	Name	Colour	Solubility in water	Revision points
NaOH	sodium hydroxide	white	soluble	Solid is deliquescent
KOH	potassium hydroxide	white	soluble	Solid is deliquescent
$Ca(OH)_2$	calcium hydroxide	white	slightly soluble	Solid often called slaked lime, solution lime water
$Mg(OH)_2$	magnesium hydroxide	white	insoluble	Precipitated by any alkali even if excess is used
$Fe(OH)_2$	iron(II) hydroxide	dirty green	insoluble	
$Fe(OH)_3$	iron(III) hydroxide	brown	insoluble	
$Al(OH)_3$	aluminium hydroxide	white	insoluble	Precipitated by using any alkali, but redissolve if the alkali is NaOH or KOH and excess is used, because the solids are amphoteric
$Pb(OH)_2$	lead(II) hydroxide	white	insoluble	
$Cu(OH)_2$	copper(II) hydroxide	blue	insoluble	Precipitated by any alkali but redissolve if the alkali is ammonia and excess is used because a soluble complex salt is formed
$Zn(OH)_2$	zinc hydroxide	white	insoluble	Precipitated by any alkali but redissolves if an excess of *any* alkali is used as the solid is amphoteric and also forms soluble complexes with ammonia

taining the appropriate metal ion (use a solution of the nitrate if in doubt, because all nitrates are soluble), e.g.

$$Cu^{2+}(aq) + 2OH^-(aq) \rightarrow Cu(OH)_2(s)$$

Unfortunately there are complications in making the hydroxides in this way, except for those of magnesium and iron. The hydroxides of the other metals must be prepared by adding the solution of OH^- ions (alkali) slowly, drop by drop, to the metal ion solution until precipitation is just complete. If excess alkali is added, the precipitate first formed may then dissolve. The actual effect depends on both the individual alkali and the metal ion, and these and other factors are summarized in Table 8.3.

Properties of metallic hydroxides
1 The only general property of importance is their ability to act as bases, and also as alkalis if soluble in water.
2 Some metal hydroxides decompose to the oxide when heated (see preparation of oxides, p. 115).
3 The precipitation of a metal hydroxide from solution is a useful test for a particular metal ion. The colour of the precipitate and/or its behaviour when excess alkali is added (see Table 8.3 and the tests on p. 170) often indicate which metal hydroxide has been formed, and therefore which metal ion is present.

8.3 Metallic carbonates

Methods of preparation
Precipitation is the only important general method of preparing carbonates. All metal carbonates are insoluble except for those of sodium and potassium, solutions of which are used to provide the $CO_3^{2-}(aq)$ ions needed for the precipitation of the others, e.g.

$$CO_3^{2-}(aq) + Mg^{2+}(aq) \rightarrow MgCO_3(s)$$

As with all precipitations, remember that the metal ion must be provided in a *solution*; use the nitrate if you are not sure about the solubilities of any other compounds.

The only (minor) complication is that some of the precipitates will not be pure carbonates, for they precipitate as a mixture of the carbonate and the hydroxide; such solids are sometimes called *basic carbonates*.

Remember that the carbonates of iron(III) and aluminium do not exist. See Table 8.4 for a general summary of carbonates.

Table 8.4 *The common metallic carbonates*

Formula	Name	Solubility in water	Usual appearance	Revision points
Na_2CO_3	sodium carbonate	soluble	colourless crystals or white powder	The crystals, $Na_2CO_3.10H_2O$, effloresce to a white powder, the monohydrate
$CaCO_3$	calcium carbonate	insoluble	white powder	Occurs in nature as limestone, marble, etc.
$MgCO_3$	magnesium carbonate	insoluble	white powder	–
$ZnCO_3$	zinc carbonate	insoluble	white powder	–
–	aluminium carbonate	–	–	Does not exist
$FeCO_3$	iron(II) carbonate	insoluble	green powder	–
–	iron(III) carbonate	–	–	Does not exist
$PbCO_3$	lead(II) carbonate	insoluble	white powder	–
$CuCO_3$	copper(II) carbonate	insoluble	green powder	–

Properties of metallic carbonates

1 Carbonates of some metals decompose to the oxide and carbon dioxide on heating (see preparation of oxides, p. 114).
2 Carbonates (and hydrogencarbonates) cause effervescence when added to dilute acids, forming a salt, water and carbon dioxide. This reaction is used as a standard salt preparation (p. 74), as well as providing a test for a carbonate (or hydrogencarbonate) and a laboratory preparation of carbon dioxide (p. 146).
3 The only hydrogencarbonate of any importance is $NaHCO_3$, and as its properties are very similar to those of Na_2CO_3 it is useful to be able to distinguish between them. Sodium hydrogencarbonate (in solid form or in solution) will liberate carbon dioxide on heating, but sodium carbonate will not do so.

$$2NaHCO_3(s) \rightarrow Na_2CO_3(s) + CO_2(g) + H_2O(g).$$

8.4 Metallic nitrates

Methods of preparation
All nitrates are soluble, and all nitrates are salts, so they are prepared by one of the methods used for preparing salts (Unit 4.2), i.e. by the action of nitric acid on the appropriate metal or compound. The only thing to remember is that nitric acid is also a powerful oxidizing agent, and so if a compound of a metal with more than one valency is used, the solution formed will contain the nitrate of the metal in its higher valency. Thus the action of nitric acid on iron(II) hydroxide will produce some iron(III) nitrate. For similar reasons, dilute nitric acid will not attack aluminium metal, for the protective oxide layer on the metal is made even more effective by the oxidizing acid.

Properties of metallic nitrates
These are also summarized in Table 8.5.

Table 8.5 *Some common metallic nitrates*

Formula	Name	Solubility in water	Usual appearance	Revision points
$NaNO_3$	sodium nitrate		white crystals	When heated, decomposes to the nitrite and oxygen
$Ca(NO_3)_2$	calcium nitrate		white crystals	
$Mg(NO_3)_2$	magnesium nitrate		white crystals	
$ZnCO_3$	zinc nitrate	All Soluble	white crystals	
$Al(NO_3)_3$	aluminium nitrate		white crystals	
$Fe(NO_3)_2$	iron(II) nitrate		usually only encountered in solution	Not prepared by usual salt preparations, which produce iron(III) salts
$Fe(NO_3)_3$	iron(III) nitrate			See above
$Pb(NO_3)_2$	lead(II) nitrate		white crystals	No water of crystallization
$Cu(NO_3)_2$	copper(II) nitrate		green crystals	

(a) How to test for a nitrate You will only need to know one of the following tests.

(*1*) *The brown ring test* Place a small volume of a solution of the suspected nitrate in a boiling tube, and acidify with dilute sulphuric acid. Mix in a small volume of freshly prepared iron(II) sulphate solution, and then carefully pour a little concentrated sulphuric acid down the inside of the tube so as to form an undisturbed lower layer of liquid. Nitrates cause a brown layer at the junction of the two layers of liquid.

(*2*) *Using Devarda's alloy* Place a small volume of a solution of the suspected nitrate in a boiling tube and add an equal volume of dilute sodium hydroxide solution, followed by about 0.5 g of Devarda's alloy. Warm the tube gently until the contents boil. If ammonia gas is liberated (detected by turning damp red litmus paper blue), the solution contains a nitrate.

(b) Heating metal nitrates This has been considered on p. 114 as a method of preparing oxides. The reaction is also used to prepare nitrogen dioxide gas, in which case the particular nitrate used is lead nitrate. This nitrate does not have any water of crystallization, unlike the others, and therefore there is no water produced which could dissolve the nitrogen dioxide.

(c) Solubility It is very important to remember that all nitrates are soluble, and this has been referred to on several occasions when solutions are mixed to form precipitates. For example, nearly all lead compounds are insoluble in water, and lead(II) nitrate is one of the few compounds which can provide lead(II) ions in solution.

8.5 Metallic sulphates

Methods of preparation

1 Sulphates are salts, and so the soluble ones can be prepared by the action of dilute sulphuric acid on a suitable metal or compound as described in Unit 4.2. Of the common metallic sulphates, only $BaSO_4$, $PbSO_4$ and (the slightly soluble) $CaSO_4$ are insoluble, and these are prepared by precipitation.

2 The insoluble sulphates referred to above are prepared by adding a *solution* containing the appropriate metal ion to a *solution* containing sulphate ions (e.g. sodium sulphate solution), and then filtering and washing the precipitate formed, e.g.

$$SO_4^{2-}(aq) + Pb^{2+}(aq) \rightarrow PbSO_4(s)$$

3 Sulphuric acid is a dibasic acid and can be used to prepare hydrogensulphates as well as the normal sulphates. A typical method has been described on p. 74.

Properties of metallic sulphates (see also Table 8.6)

The test for a soluble sulphate To a small volume of a solution of the suspected sulphate, add a little dilute hydro*chloric* acid followed by some barium *chloride* solution. A white precipitate (of barium sulphate) confirms the presence of sulphate ions,

$$Ba^{2+}(aq) + SO_4^{2-}(aq) \rightarrow BaSO_4(s)$$

Table 8.6 *Some common metallic sulphates*

Formula	Name	Solubility in water	Usual appearance	Revision points
Na_2SO_4	sodium sulphate	soluble	white crystals	
$CaSO_4$	calcium sulphate	insoluble	white powder	Occurs naturally as gypsum and anhydrite
$MgSO_4$	magnesium sulphate	soluble	colourless crystals	Commonly called Epsom salts
$ZnSO_4$	zinc sulphate	soluble	white crystals	
$FeSO_4$	iron(II) sulphate	soluble	green crystals	Decomposes in an unusual way when heated, forming red iron(III) oxide (change of valency) and two different oxides of sulphur. $2FeSO_4(s) \rightarrow Fe_2O_3(s) + SO_2(g) + SO_3(g)$
$CuSO_4$	copper(II) sulphate	soluble	blue crystals	When heated loses water of crystallization to form white powder. The reverse colour change sometimes used to detect the presence of (not necessarily pure) water
$PbSO_4$	lead(II) sulphate	insoluble	white powder	
$BaSO_4$	barium sulphate	insoluble	white powder	Normally only seen as the precipitate in a positive sulphate test

Note: Many students confuse the chemicals used in the sulphate test with those used in the chloride test, particularly as the result, if positive, is a white precipitate in each case. To avoid this problem, remember that the reagents to be added go in 'matching pairs' (i.e. hydro*chloric* acid and barium *chloride*, or *nitric* acid and silver *nitrate*) and that you must never add extra chloride ions (i.e. barium chloride and hydrochloric acid) when doing a chloride test.

8.6 Metallic chlorides

Methods of preparation
1 All metallic chlorides are salts, and most of them are soluble in water. They are prepared by the typical salt preparations discussed in Unit 4.2, using dilute hydrochloric acid on an appropriate metal or compound. Only two common chlorides are insoluble, those of lead and silver.
2 The insoluble metallic chlorides referred to above are made by precipitation. A *solution* containing the appropriate metal ion is added to a *solution* containing chloride ions (e.g. sodium chloride). The resultant precipitate is filtered, washed and dried, e.g

$$Pb^{2+}(aq) + 2Cl^-(aq) \rightarrow PbCl_2(s)$$

3 The *anhydrous* chlorides of iron(II), iron(III) and aluminium have to be prepared by dry methods as described on p. 76.

Properties of metallic chlorides (see also Table 8.7)
(a) How to test for a soluble chloride To a small volume of a solution of the suspected chloride, add a little dilute *nitric* acid followed by some silver *nitrate* solution. A white precipitate (of silver chloride) confirms the presence of chloride ions,

$$Ag^+(aq) + Cl^-(aq) \rightarrow AgCl(s).$$

(b) The reaction between a solid chloride and concentrated sulphuric acid If concentrated sulphuric acid is added to a solid metallic chloride and then gently warmed, steamy fumes of hydrogen chloride gas are evolved and the mixture often froths, e.g.

$$NaCl(s) + H_2SO_4(l) \rightarrow NaHSO_4(aq) + HCl(g)$$

The presence of hydrogen chloride can be shown by allowing ammonia fumes to mix with the evolved gas, when a dense white cloud of ammonium chloride is formed,

$$NH_3(g) + HCl(g) \rightarrow NH_4Cl(s)$$

This reaction is used in the laboratory preparation of hydrogen chloride (p. 163) and as a test for an *insoluble* chloride.

Table 8.7 *Some common metallic chlorides*

Formula	Name	Solubility in water	Usual appearance (i.e. hydrated)	Revision points
NaCl	sodium chloride	soluble	white crystals	
$CaCl_2$	calcium chloride	soluble	white crystals	Anhydrous salt often used as a drying agent, but not for ammonia
$MgCl_2$	magnesium chloride	soluble	white crystals	
$ZnCl_2$	zinc chloride	soluble	white crystals	
$AlCl_3$	aluminium chloride	soluble	white crystals	Anhydrous salt must be made by dry methods
$FeCl_2$	iron(II) chloride	soluble	not often used	Anhydrous salt made as above
$FeCl_3$	iron(III) chloride	soluble	brown solid	Anhydrous salt made as for aluminium
$PbCl_2$	lead(II) chloride	insoluble	white crystals	Soluble in hot water
CuCl	copper(I) chloride	insoluble	white crystals	Soluble in ammonia solution and conc. hydrochloric acid due to formation of complexes
$CuCl_2$	copper(II) chloride	soluble	blue-green crystals	
AgCl	silver chloride	insoluble	white solid	Normally only seen as the precipitate in a positive chloride test

If the reaction is repeated but with the addition of some manganese(IV) oxide, MnO_2 (i.e. the reaction mixture consists of a solid chloride, solid manganese(IV) oxide and concentrated sulphuric acid), the hydrogen chloride first formed is immediately oxidized to chlorine by the manganese(IV) oxide. Chlorine gas is thus the main product.

$$NaCl(s) + H_2SO_4(l) \rightarrow NaHSO_4(aq) + HCl(g)$$
and then
$$2HCl(g) + [O] \rightarrow H_2O(l) + Cl_2(g)$$

8.7 Sodium and potassium compounds

The metals
1 They are very reactive (e.g. at the top of the activity series) because their atoms need to lose only one electron to attain a stable structure.

2 They are very soft and have low densities (float on water).
3 They react with oxygen in the air, and the compounds formed then react with other compounds in the air. A typical series of reactions is:

$$Na(s) \xrightarrow{O_2(g)} Na_2O(s) \xrightarrow[\text{with water vapour}]{\text{deliquesces (p. 134), reacts}} NaOH(aq)$$

$$\text{surface coating of } Na_2CO_3.H_2O \xleftarrow{\text{effloresces (p. 134)}} Na_2CO_3.10H_2O(s) \xleftarrow{CO_2(g)}$$

To avoid these changes the metals are stored under a liquid hydrocarbon.

4 They react with cold water to form an alkaline solution and hydrogen gas, e.g.

$$2Na(s) + 2H_2O(l) \rightarrow 2NaOH(aq) + H_2(g)$$

5 Other reactions of the metals (e.g. combustion in oxygen, reaction with chlorine) are considered under appropriate headings elsewhere in the book.

The compounds of sodium and potassium

1 All sodium and potassium compounds are soluble in water. Whenever you need to provide anions in solution (e.g. SO_4^{2-}), the corresponding sodium or potassium salts will *always* be satisfactory.
2 All sodium and potassium compounds are ionic.
3 All sodium and potassium compounds are white unless the compound also contains a transition element. The only common coloured compounds of sodium and potassium are their orange dichromate(VI) compounds and the purple potassium manganate(VII).
4 Tests for sodium and potassium ions are given on p. 170.

Some uses of sodium and its compounds

The metal: in street lamps, and as a coolant in nuclear reactors.
Sodium hydroxide: manufacture of paper, artificial silk and soap.
Sodium carbonate: glass and paper making.
Sodium hydrogencarbonate: baking powder.
Sodium chloride: food industry, glazing pottery and the manufacture of sodium, chlorine and sodium hydroxide.

8.8 Calcium, magnesium and zinc compounds

Calcium metal
1 This is a reactive metal (near the top of the activity series) as it easily loses two electrons to form a stable ion.
2 It is sometimes stored under a hydrocarbon oil, for it reacts with oxygen in the air and then with other substances, e.g.

$$Ca(s) \xrightarrow{O_2(g)} CaO(s) \xrightarrow{H_2O \text{ vapour}} Ca(OH)_2(s) \xrightarrow{CO_2(g)} CaCO_3(s)$$

3 It reacts fairly vigorously with *cold* water to form an alkaline solution of calcium hydroxide (lime water),

$$Ca(s) + 2H_2O(l) \to Ca(OH)_2(aq) + H_2(g)$$

Magnesium and zinc metals
It is likely that you have studied the properties of these fairly reactive metals in detail, e.g. on steam (p. 58), dilute acids (p. 61), oxygen (p. 58) and displacement reactions with salts of less reactive metals (p. 59).

The compounds of calcium, magnesium and zinc
These metals always have a valency of two, so that if you know the formulae of the compounds of one of them, the formulae of the corresponding compounds of the other metals are easily determined.

1 All the common compounds of these metals are white and ionic.
2 The tests for these metals are given on p. 170.
3 Some syllabuses require you to be familiar with some of the chemistry of the following calcium compounds, which have not been included in Tables 8.1 to 8.7.

Calcium carbide, CaC_2 This is prepared industrially by the reaction between calcium oxide and coke in an electric furnace,

$$CaO(s) + 3C(s) \to CaC_2(s) + CO(g)$$

The carbide is used to make calcium cyanamide, which has important applications as a fertilizer and in the manufacture of melamine plastics and synthetic resins.

Calcium carbide produces ethyne (acetylene) when water is added,

$$CaC_2(s) + 2H_2O(l) \to Ca(OH)_2(aq) + C_2H_2(g)$$

This reaction is likely to become increasingly important as a source of hydrocarbons as the World's supply of crude oil declines.

Calcium phosphate, $Ca_3(PO_4)_2$ The impure salt is mined as rock phosphate, and is converted into phosphorus, phosphates and fertilizers (e.g. superphosphate).

Bleaching powder This is made by treating calcium hydroxide with chlorine and the essential chemical in it is calcium chlorate(I). Bleaching powder is a useful source of chlorine, as well as being a bleaching agent. Most liquid bleaches contain a chlorate(I) compound (hypochlorite), which readily yields chlorine for bleaching purposes.

Some uses of magnesium, calcium, zinc and their compounds

The metals: magnesium is present in a range of important low density alloys; zinc is used in alloys (e.g. brass, which also contains copper), for protecting iron from corrosion by galvanizing, and in dry cells.

Calcium carbonate: in the blast furnace; to prepare the oxide and hydroxide, and as marble in building.

Calcium hydroxide: in the manufacture of mortar and cement, as a cheap industrial base, for neutralizing acid soils and for softening temporarily hard water.

Calcium sulphate: gypsum is used to prepare Plaster of Paris, $(CaSO_4)_2H_2O$, plaster and plasterboard.

Magnesium compounds: sometimes used in medicines, e.g. a suspension of the hydroxide is the active ingredient of 'milk of magnesia' (to neutralize acidity) and the sulphate is a laxative (Epsom salts).

Zinc compounds: are used in paints and ointments, e.g. the carbonate in calamine lotion.

8.9 Aluminium and its compounds

The metal
1 The metal does not normally show its reactive nature because it is protected by a thin but very tenacious film of oxide.
2 If a piece of aluminium is made the anode of a cell in which oxygen is liberated by electrolysis, the oxygen reacts with the aluminium to thicken the thin oxide film already present. This process is called **anodizing**. The thicker layer of oxide protects the metal even more than usual and readily adsorbs a dye. Materials

made of anodized aluminium (e.g. saucepans) are thus highly resistant to corrosion and can be attractively coloured.

3 The other reactions of aluminium with which you should be familiar have been referred to elsewhere in the book, its action on acids (p. 61), on sodium hydroxide (p. 70) and on chlorine (p. 75).

Aluminium compounds

1 All the common ones are white.
2 It is quite difficult for an aluminium atom to lose three electrons in order to form a stable ion, and so you cannot assume that its compounds are always ionic.
3 The test for aluminium compounds is given on p. 171.

Some uses of aluminium

The metal is particularly useful because of its lightness, non-toxicity and resistance to corrosion. For one or more of these reasons it is used to make products such as saucepans, storage tanks and milk churns, and also as a foil for packaging. The metal is also often alloyed and used in aircraft bodies, 'aluminium' framed greenhouses and other materials where lightness and resistance to corrosion is required.

8.10 Iron and its compounds

The metal

Iron is the most common metal in the world, and you will be familiar with many of its reactions. Most of these are considered elsewhere in the book, e.g. its action on acids (p. 61), on steam (p. 58) and on chlorine (p. 75).

Rusting

When iron is exposed to both oxygen and moisture it rusts, and this is one of the biggest disadvantages of iron and steel. (Rust is mainly hydrated iron(III) oxide.) You could be asked to describe an experiment which shows the factors responsible for the rusting of iron. A typical example involves exposing iron nails to different environments in test-tubes.

Iron and (particularly) steel are used in spite of the fact that they rust so easily, because they are also relatively cheap to produce. Nevertheless, the cost of combating corrosion in Britain alone runs into millions of pounds each year. The methods used against corrosion usually work by ensuring that both oxygen and water are unable to come into contact with the metal, e.g. painting, tin-plating, greasing, galvanizing and coating with plastics.

Another method of combating corrosion is of an electrical nature, in which a smaller piece of metal more reactive than iron (e.g. magnesium) is bolted to the iron or steel (e.g. the hull of a ship) and acts as a 'sacrificial anode'. In the presence of an electrolyte (e.g. sea water) a cell is set up in which electrons leave the atoms of the more reactive metal and flow to the iron. The more reactive metal 'dissolves' as its ions are formed, but the iron does not change. A reactive metal used in this way can be renewed regularly.

Types of iron

Iron from the blast furnace is very impure, and is usually processed further. It is useful to distinguish between the properties of cast iron, wrought iron and steel (see Table 8.8).

Table 8.8 *Cast iron, wrought iron and steel*

Cast iron	Wrought iron	Steel
Contains about 4% carbon with a small percentage of silicon, phosphorus and sulphur. This alloy of iron expands on solidification. It is brittle and is used where strength is not important.	Much purer form obtained by heating cast iron with iron(III) oxide. This is a softer form and can be worked, e.g. made into chains. It is purer because it contains less sulphur, phosphorus and silicon, but it still contains carbon.	This is an alloy of iron containing between 0.15 and 1.5% carbon. Its properties can be altered in several ways: (a) by altering the percentage of carbon, (b) by heat treatment, (c) by adding other metals such as chromium or manganese. The use of steel is limited only by its high density and tendency to corrode.

Iron compounds

(a) All the iron compounds you will study are coloured, and so it is important to learn the colours of the individual compounds. Remember that there are iron(II) and iron(III) compounds.

(b) The oxidation of an iron(II) compound to an iron(III) compound
Note: Oxygen in the air slowly oxidizes iron(II) compounds to iron(III) compounds, so samples from opened bottles are rarely pure.

This reaction is an oxidation because the iron(II) ions lose electrons,

$$Fe^{2+}(aq) - e \rightarrow Fe^{3+}(aq)$$

Many oxidizing agents will bring about this change, e.g. hydrogen peroxide, concentrated nitric acid, acidified potassium manganate(VII), acidified potassium dichromate(VI), concentrated sulphuric acid and chlorine. Wherever possible, try not to add any 'new' anions during the oxidation, e.g. chlorine would be used to oxidize iron(II) chloride to iron(III) chloride. Concentrated sulphuric acid would also achieve this, but would also provide sulphate ions so that if the resultant solution were crystallized, a mixture of iron(III) sulphate and iron(III) chloride could be produced.

(c) The reduction of an iron(III) compound to an iron(II) compound

The best reducing agent to use is a combination of iron metal and a dilute acid containing the same anion as that in the compound. The mixture must be warmed. For example, to reduce iron(III) chloride to iron(II) chloride, add iron filings and warm dilute hydrochloric acid. To reduce iron(III) sulphate to iron(II) sulphate, add iron filings and warm dilute sulphuric acid. In each case, the metal reacts with the acid, liberating electrons

$$Fe(s) \rightarrow Fe^{2+}(aq) + 2e$$

Some of these electrons combine with hydrogen ions from the acid to liberate hydrogen gas,

$$2H^{+}(aq) + 2e \rightarrow H_2(g)$$

and others reduce the iron(III) ions already present,

$$2Fe^{3+}(aq) + 2e \rightarrow 2Fe^{2+}(aq)$$

(d) The tests which distinguish between iron(II) and iron(III) compounds are given on p. 171.

8.11 Lead, copper and their compounds

Lead metal

Lead is a relatively unreactive metal, and its chemical properties are rarely studied at an elementary level. The freshly cut metal is shiny, but it rapidly tarnishes in air due to the formation of the basic carbonate, which then prevents further corrosion. Lead is very dense but also very soft.

Lead compounds
The only lead compounds normally studied at an elementary level are the lead(II) compounds and the various oxides. The oxides are a little unusual (see Table 8.2). The only common soluble lead(II) salt is lead(II) nitrate. The test for lead ions is given on p. 171.

Copper metal
This is also relatively unreactive. The only chemical reaction of any importance is that with dilute nitric acid, for this is the only convenient way of turning the metal into a solution of its ions.

Copper compounds
The metal forms copper(I) compounds and copper(II) compounds, but only the latter are considered in most courses. (Some syllabuses require you to know a little about the chemistry of copper(I) oxide and copper(I) chloride, and so these have been included in Tables 8.2 and 8.7.)

All copper(II) compounds are coloured. They show no exceptional behaviour in their reactions. The test for copper(II) ions is given on p. 170.

Some uses of copper and its compounds
The metal is much used in plumbing, for it is resistant to corrosion and fairly easy to work with. It is also used to make electrical wiring as it resists corrosion and is an excellent conductor of electricity. Common alloys containing the metal include the bronzes (copper and tin), the brasses (copper and zinc) and the coinage alloys (copper, tin and zinc for pennies, etc. and copper and nickel for 'silver' coins). The attractive appearance of the pure metal is also utilized in decorative and ornamental work.

Copper(I) oxide is used in making red glass, and copper(II) sulphate is used in garden fungicides and in electroplating.

Questions

1 What type of reaction is represented by the following equation?
$$Fe^{2+} \rightarrow Fe^{3+} + e^-$$
Name *two* reagents which could be used separately to bring about this reaction. What reagent would you use to test that this reaction had taken place? What would you expect to observe during this test? (JMB)

2 Sodium metal is kept under: (*a*) water; (*b*) ethanol; (*c*) nitric acid; (*d*) paraffin oil; (*e*) mercury.

3 Iron is galvanized by coating it with: (a) copper; (b) tin; (c) zinc; (d) aluminium; (e) lead.

4 A white crystalline compound exposed to air changed into a white powder. The crystalline compound may have been: (a) potassium hydroxide; (b) calcium oxide; (c) sodium nitrate; (d) sodium carbonate; (e) potassium sulphate.

5 A mixture consists of a soluble salt A and an insoluble salt B. When the mixture is heated it turns black and a colourless gas which turns calcium hydroxide solution milky is evolved.

If the mixture is shaken with water and filtered, it gives a colourless solution and leaves a green residue.

The solution gives the following results on testing: (a) with sodium carbonate solution there is a white precipitate; (b) a flame test gives a brick-red coloured flame; (c) when dilute nitric acid is added followed by silver nitrate solution, there is a curdy white precipitate.

When the residue on the filter paper is dissolved in the minimum of dilute nitric acid, there is effervescence and a green solution is formed which gives: (i) a brownish-black precipitate with hydrogen sulphide; (ii) a light blue precipitate with sodium hydroxide; (iii) a light blue precipitate with ammonium hydroxide, soluble in excess of the latter to give a deep blue solution.

Identify the salts A and B and explain all the reactions described above, giving equations. (AEB)

6 Describe the effect of: (a) strongly heating a piece of marble in a Bunsen flame; (b) moistening the product from (a) with water.

Give the common name of the product in each case.

Starting with a piece of marble, how would you prepare a pure dry sample of calcium sulphate? For what reason is slaked lime added to the soil in the garden? Why is it inadvisable to lime the soil shortly after applying ammonium sulphate as a fertilizer? (JMB)

7 Name two important ores of iron. Outline the extraction of iron by the blast furnace process, explaining the essential chemistry. (No diagram is required, nor are technical details.) What differences are there in composition between cast iron and steel? Describe, giving any necessary conditions, the action of: (a) hydrochloric acid; (b) nitric acid (one reaction only); and (c) steam on iron. (CAM)

9 Water, soaps and detergents. Hydrogen peroxide

9.1 Water as a chemical. Water pollution

Water as a chemical

We tend to forget that water is a chemical, and that we study many of its reactions in chemistry, such as the reaction of water or steam on metals (p. 58). Hydrolysis, hydration, dehydration and drying are terms frequently misused in referring to chemical reactions involving water.

*When water reacts chemically with a compound to form at least two products, the reaction is called a **hydrolysis**.*

In elementary work you are likely to use this term only in describing the breakdown of condensation polymers such as starch. Such molecules are formed when molecules (the monomers) join together by a chemical reaction which also produces water (p. 184); in the reverse process, water reacts with (hydrolyses) the polymer to reform the original monomers.

*When water combines with a substance to form one product only, in such a way that water molecules remain intact but bonded to the product the substance is said to be **hydrated**.*

Notice that in hydration the water is actually *combined* with the substance, e.g. as water of crystallization. A hydrated solid may be perfectly dry, even though it contains water. For this reason, dehydration (a chemical reaction which removes *combined* water) is not the same as drying.

Many metal salts contain water of crystallization, and it is important to remember that an anhydrous salt and its corresponding hydrated salt are often quite different in appearance, i.e. colour, and size and shape of crystals. For example, hydrated copper(II) sulphate consists of blue crystals, whereas the anhydrous salt appears as a white powder (the crystals are too small to be distinguished with the naked eye). Similarly, sodium carbonate deccahydrate ($Na_2CO_3.10H_2O$) has large, colourless crystals, whereas the monohydrate appears as a white powder.

*A **deliquescent** substance is one which absorbs water vapour from the atmosphere and then dissolves in it to form a solution, e.g. sodium hydroxide.*

*An **efflorescent** substance is one which loses some or all of its water of crystallization when exposed to the atmosphere. Sodium carbonate deccahydrate effloresces to the monohydrate.*

*A **hygroscopic** substance absorbs water vapour from the air and becomes damp, but does not dissolve in the water to form a solution, e.g. copper(II) oxide.*

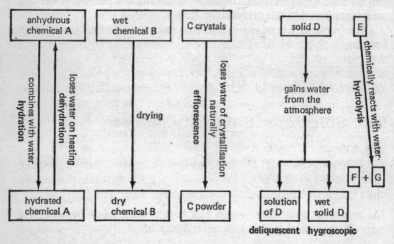

Water pollution

Water is often described as the universal solvent because it dissolves most of the substances familiar to us, even if only to a small extent. We tend to forget that most of the chemical experiments we conduct take place in aqueous solution, and of course this is also true of many industrial processes. Similarly, some of the waste products from our bodies are removed in aqueous solution, and much of the waste from chemical factories is removed in aqueous solution. There is increasing concern about the materials which find their way into the streams, rivers and seas and which pollute them. It is becoming increasingly difficult to purify water for domestic use and to ensure that water supplies are kept free from pollution.

The major pollutants in the rivers and seas include nitrates, phosphates, detergents and ions of heavy metals such as mercury. Nitrates are used on a vast scale as fertilizers, and as they are very soluble they are sometimes found in streams draining agricultural land. Phosphates are also washed into streams from excess fertilizers. Nitrates and phosphates cause problems in rivers and lakes, and even low concentrations in drinking water constitute a health hazard.

Detergents are used on an enormous scale and get washed into rivers. Some of the early synthetic detergents could not be broken down by bacteria, and so they persisted in rivers, causing a blanket of foam and having harmful effects on fish and other forms of life. Most detergents are now 'biodegradeable', i.e. they can be broken down by bacteria, but even so their use on a large scale (e.g. to disperse oil slicks at sea) can have undesirable side effects on fish and other forms of marine life.

Large quantities of mercury are used in the Kellner-Solvay process (p. 56), and small amounts have escaped as mercury ions in the waste materials. The problem here is that even very small traces of mercury ions (and other heavy metal ions) can cause serious illness, and once the ions are in the body they are very difficult to remove.

9.2 Solutions. Solubility curves

Solubility

If a soluble solid is added in small quantities at a time to a fixed volume of water at a fixed temperature, a point is reached when no more will dissolve, and a saturated solution is formed.

*The **solubility** of a solid in water at a particular temperature is the mass of solid which will dissolve in 100 g of water at that temperature.*

To ensure that the solution is saturated, measurements of solubility are usually taken when excess undissolved solid is in contact with the solution.

A supersaturated solution is one which contains more dissolved solid than would be expected from its solubility at that temperature. Such solutions are unstable; the addition of even the slightest trace of solid (e.g. a crystal of the solute, or even dust), or shaking, will ensure that dissolved solid comes out of solution. A common example of a supersaturated solution is that produced by gently warming sodium thiosulphate crystals, when they dissolve in their own water of crystallization. If the solution is allowed to cool and one crystal is added, a mass of crystals forms rapidly with a considerable rise in temperature.

To determine the solubility of a solid in water

If you have done an experiment of this kind, you could be asked to describe it in detail, or to use given results from such an experiment to calculate a solubility. Some typical experimental results might be used as follows.

Temperature of solution: 25 °C
Mass of evaporating basin + 20 cm³ sample of saturated solution: 46.2 g
Mass of empty evaporating basin: 25.0 g
∴ mass of 20 cm³ sample of saturated solution: 21.2 g
Mass of evaporating basin + solid after careful evaporation of solvent: 26.2 g
∴ mass of solid dissolved in 20 cm³ = (26.2 − 25.0) g = 1.2 g
1.2 g solid dissolve in 20 cm³ of water at 25 °C
∴ 6.0 g solid dissolve in 100 g (100 cm³) of water at 25 °C, and this is the solubility of the solid at 25 °C.

Note: The temperature must always be stated, because the solubility of most solids increases with a rise in temperature.

Solubility curves

In order to show how the solubility of a solid varies with temperature, the data shown above must be obtained for solutions at

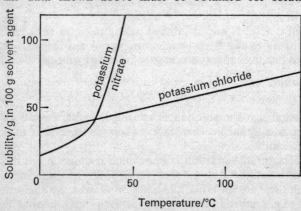

The solubility curves for potassium nitrate and potassium chloride in water

Water, soaps and detergents. Hydrogen peroxide 137

different temperatures. The results are often plotted on a graph such as that shown below, when they are known as a **solubility curve**. You could be asked to use some experimental results in order to produce a solubility curve, or you could be asked to describe an experiment to determine the solubility curve for a solid.

Some uses of solubility curves
1 It is possible to determine the mass of crystals which would be obtained by cooling a volume of the hot saturated solution from one known temperature to another.
2 One substance can be separated from another by using the fact that they have different solubilities in a solvent. It is possible to make a hot solution of a mixture of two solids (A and B) so that it is saturated with respect to one of the substances (A) but not the other (B). On cooling the solution, A can no longer remain dissolved and starts to crystallize, but B remains in solution. This technique is called **fractional crystallization**, and it is frequently used to separate mixtures and to purify a substance from soluble impurities.

9.3 Soaps and synthetic detergents

Soapy detergents (traditional soaps) These are made by hydrolysing a natural fat with sodium (or potassium) hydroxide solution, when one of the products is the sodium (or potassium) salt of an organic acid such as stearic acid, oleic acid or palmitic acid. Such acids have fairly large molecules containing about 18 carbon atoms, and their sodium (or potassium) salts are soaps, e.g. sodium stearate. Perfume, colouring and other substances may be added, and the product may be sold as soft soaps, hard soaps, soap flakes or toilet soaps. Soapy detergents are now 'outnumbered' in use by the soapless or synthetic detergents. Soaps are made from potential foods (animal fat or vegetable oil) and are limited in use, e.g. they are rendered momentarily useless by hard water, high and low pH, and cannot be used in sea water (which contains Ca^{2+}(aq) and Mg^{2+}(aq) ions). Synthetic detergents do not have these disadvantages.

Synthetic (soapless) detergents These have the same essential structure as a soap (see *How a detergent works*), but they are made by taking hydrocarbons from crude oil and substituting in them an ionic group, e.g. $-O-SO_3^-$ or $\langle O \rangle-SO_3^-$, where $\langle O \rangle$ is benzene.

They have another advantage over soaps (i.e. in addition to those given above) in that many varieties are possible, so that a synthetic detergent can be 'tailor made' for a particular job such as dispersing an oil slick at sea. Some modern detergents also contain enzymes, which help to break down stains which are stubborn towards the detergent itself.

How a detergent works

When a detergent is added to water, it dissolves as two separate ions, e.g. Na^+ ions and stearate$^-$ ions. The anions (e.g. the stearate ions) are the effective cleaning agent.

The anion of a detergent consists of a relatively long hydrocarbon chain (which is covalently bonded) at the end of which is a small ionic group. The covalent chain is hydrophobic, i.e. water hating (see properties of covalent compounds, p. 31), whereas the ionic 'head' is hydrophilic, i.e. water loving. These are the essential features of the active part of any detergent.

How a detergent lowers the surface tension of water

Water on its own is not a good 'wetting agent' because it does not readily spread completely over a surface or penetrate the fibres of fabrics. This is mainly due to the relatively large intermolecular forces between water molecules which give it a high surface tension and tend to prevent it spreading easily. A detergent lowers the surface tension of water by reducing the intermolecular forces, and makes water a better wetting agent.

(a) Droplet of water on a surface. The surface is not 'wetted' efficiently.

(b) The droplet spreads when a detergent is added because the water becomes a better wetting agent. Molecules of water in the droplet are dispersed because the hydrophobic head of each detergent molecule is attracted to a water molecule, and the hydrophobic tail repels other water molecules.

hydrophobic tail
repels other
water molecules

hydrophilic head
attracted to
water molecules

(c) This effect lowers the surface tension of water and makes it a better wetting agent.

How a detergent lowers the surface tension of water

This happens because when a detergent is added to water, the hydrophilic groups attract water molecules but the hydrophobic groups repel them. The effect is to reduce the intermolecular forces, as shown in the Figure.

Removing dirt and grease

Grease is usually organic and is covalently bonded. The hydrophobic chains of the detergent molecules readily 'dissolve' in the grease molecules as they, too, are covalent. The heads of the detergent molecules dissolve in water molecules. The result is that the grease is broken down into smaller fragments and that two normally immiscible substances (grease and water) are brought together, so that the grease can be washed away with the water. Dirt is removed at the same time because it usually adheres to the grease.

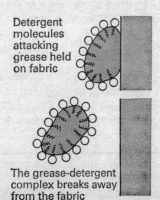

Detergent molecules attacking grease held on fabric

The grease-detergent complex breaks away from the fabric

The action of detergents on grease (simplified). The ionic heads are carried away with the water and take the grease with them

9.4 Hardness of water

Drinking water, even after purification, still contains *dissolved* substances. In normal circumstances these are harmless to health, and indeed some are beneficial, but dissolved calcium and magnesium salts cause the water to be hard, and this is inconvenient.

Hard water is water which does not readily form a lather with soap.

In Britain calcium salts are mainly responsible for hardness of water. There are two mains ways in which calcium salts get into natural water.
1 Gypsum and anhydrite ($CaSO_4.2H_2O$ and $CaSO_4$ respectively) occur naturally in Britain. Calcium sulphate is only slightly soluble in water, but streams flowing over it dissolve enough of the mineral to make the water hard.
2 Calcium carbonate (which occurs extensively in Britain as limestone and chalk) is insoluble in water, but rain water containing dissolved carbon dioxide *reacts* with it to form soluble calcium hydrogencarbonate, which dissolves and is carried away in streams and rivers.

$$CaCO_3(s) + H_2O(l) + CO_2(aq) \rightarrow Ca(HCO_3)_2(aq)$$

Note: Only *dissolved* calcium (and magnesium) compounds make water hard, and the essential factors responsible for the hardness are the calcium (and magnesium) *ions*. The anion is irrelevant.

Why hard water is inconvenient
1 When a soapy detergent dissolves in hard water, the anions (e.g. stearate ions) react immediately with the Ca^{2+}(aq) and/or Mg^{2+}(aq) ions to form a precipitate (e.g. magnesium or calcium stearate), which is seen as a scum. This is unpleasant, and difficult to remove from sinks, baths and fabrics.
2 When a soapy detergent is added to hard water, it cannot clean or form a lather until enough has been added to precipitate all of the Ca^{2+}(aq) and Mg^{2+}(aq) ions present. Only after this point are free (e.g.) stearate ions available as cleansing agents. Hard water thus wastes traditional soaps (but not soapless detergents).
3 The reaction between calcium carbonate and rain water (containing carbon dioxide) is reversible. In nature the reaction proceeds almost entirely to the right as written below, because the product is constantly washed away and equilibrium is never established.

Water, soaps and detergents. Hydrogen peroxide 141

If hard water is boiled, however, the reaction proceeds almost entirely in the opposite direction, for this time the steam and carbon dioxide escape.

$$\underset{\textbf{insoluble}}{CaCO_3(s)} + CO_2(aq) + H_2O(l) \underset{\text{boiling}}{\overset{\text{in nature}}{\rightleftharpoons}} \underset{\textbf{soluble}}{Ca(HCO_3)_2(aq)}$$

When this type of hard water is boiled, the reaction therefore produces a deposit of insoluble calcium carbonate. This builds up inside kettles as 'kettle fur' in hard water areas, and is inconvenient, but the problem is far worse on an industrial scale where millions of gallons of water may be used for heating purposes. Pipes and boilers which contain hot, hard water thus become coated with increasingly thick layers of calcium carbonate; at the best their efficiency is greatly reduced, and at the worst they could become completely blocked with consequent risk of explosion.

Permanently hard water and temporarily hard water

Permanently hard water
Cannot be softened by boiling, although it can be softened by other methods. Usually contains dissolved calcium or magnesium sulphate only.

Temporarily hard water
Can be softened by boiling. It usually contains dissolved calcium hydrogencarbonate. On boiling, insoluble calcium carbonate is formed and thus dissolved $Ca^{2+}(aq)$ ions are removed from the water, so that it is softened.

Note: Hard water can be a mixture of temporarily hard and permanently hard water.

Methods of softening water As only dissolved $Ca^{2+}(aq)$ and/or $Mg^{2+}(aq)$ cause hardness, water can be softened by any method which either removes these ions completely from the water, or which converts the ions into insoluble compounds. The main methods are summarized in Table 9.1.

9.5 Hydrogen peroxide, H_2O_2

Physical properties It is normally used in dilute aqueous solution as a colourless liquid.

Decomposition into oxygen The substance is unstable and readily decomposes into oxygen and water:

$$2H_2O_2(aq) \rightarrow 2H_2O(l) + O_2(g)$$

Table 9.1 *Methods of softening water*

Method	Type of hardness removed	How hardness is removed	Any other points
1 Boiling	Temporary hardness only	Ca^{2+}(aq) converted into insoluble calcium carbonate (learn the equation)	Totally unsuitable on a large scale, as a great deal of energy is required, and thus uneconomic to use. Solid formed can be inconvenient
2 Addition of calcium hydroxide	Temporary hardness only	$Ca(OH)_2 + Ca(HCO_3)_2$ $\downarrow$ $2CaCO_3 + 2H_2O$ Dissolved Ca^{2+}(aq) converted into insoluble calcium carbonate	Known as Clark's method. Cheap, used on large scale at water treatment plants. Only a calculated amount must be added; excess only makes the water hard again by addition of Ca^{2+}(aq) ions
3 Addition of sodium carbonate	Both types	Dissolved ions converted into insoluble carbonates and precipitated, e.g. $Ca^{2+}(aq) + CO_3^{2-}(aq)$ $\downarrow$ $CaCO_3(s)$	Very convenient. On a large scale it is more economic to use (2) to remove temporary hardness and then this method to complete the softening
4 Ion exchange	Both types	Water flows through ion exchange resin. Sodium, potassium or hydrogen ions on resin exchange with calcium or magnesium ions in hard water; ions responsible for hardness entirely removed	Can be used on a small scale (e.g. as Permutit) in appliances which are fitted to taps. Convenient on a large scale and also used to 'deionize' water, in which case anions are also removed from the water
5 Phosphate treatment	Both types	Complex polyphosphates (e.g. Calgon) 'lock up' calcium and/or magnesium ions by forming a complex structure with them	Has additional advantage of preventing corrosion of pipes carrying water. Calgon is a component of most washing up liquids

This reaction is speeded up by adding a positive catalyst such as manganese(IV) oxide as in the preparation of oxygen (p. 84). This reaction also helps to explain why the substance is an oxidizing agent.

As an oxidizing agent If you are asked to construct an equation to show how hydrogen peroxide oxidizes something, it may be convenient to show the reaction as one involving the addition of oxygen,

$$H_2O_2(aq) \rightarrow H_2O(l) + (O) \text{ (to substance being oxidized)}$$

or as one involving the loss of one or more electrons,

$$H_2O_2(aq) + 2H^+(aq) + 2e \text{ (from substance being oxidized)} \rightarrow 2H_2O(l)$$

Note that in the second reaction the hydrogen peroxide must be acidified. Typical oxidations performed by hydrogen peroxide include the following.

1 Black lead sulphide to white lead sulphate,

$$4H_2O_2(aq) \rightarrow 4H_2O(l) + 4[O]$$
$$PbS(s) + 4[O] \rightarrow PbSO_4(s)$$

This reaction is used to restore old oil paintings, where white pigments based on lead compounds may have darkened due to formation of lead sulphide.

2 The conversion of iron(II) salts to iron(III) salts, as on p. 129.

As a bleaching agent Hydrogen peroxide is a powerful bleaching agent, and will bleach human hair, synthetic fibres and some stains. As with the bleaching action of chlorine (p. 163), these are really further examples of its oxidizing properties.

Questions

1 It is very difficult to form a lather with ordinary soap in sea-water.
  (a) Why is this so?
  (b) What would be formed instead of the lather?
  (c) What type of cleanser would form a lather with sea-water? Why is this type of substance effective when soap is not? (JMB)

2 Differentiate between 'temporarily' and 'permanently' hard water.

3 Describe carefully what happens when carbon dioxide is passed through calcium hydroxide solution (limewater) until no further change occurs. Give equations.

4 Explain the terms 'efflorescent', 'deliquescent' and 'hygroscopic', and give examples to illustrate your answer.

5 Which one of the following *cannot* be used to soften permanently hard water?
(a) Calcium hydroxide.
(b) Distillation.
(c) Sodium carbonate.
(d) An ion exchange resin.
(e) Calgon.

6 Which of the following is not true about the hydrated and anhydrous forms of copper(II) sulphate?
(a) One is blue and the other is colourless (white).
(b) One is dry and the other is wet.
(c) One is crystalline and the other is powdered.
(d) They have different formula masses.
(e) Energy changes are involved in their inter-conversions.

7 (a) Calcium carbonate, present in rocks or soil, is one of the causes of hardness of water. Explain why this is so.

(b) Explain the use of (i) calcium hydroxide; and (ii) sodium carbonate, in the softening of hard water.

(c) Why does the presence of dissolved sodium carbonate not make water hard?

(d) A copper boiler used in the preparation of distilled water is encrusted with a layer of white scale caused by the hardness in the water used. Explain how this scale was formed from the hard water.

If supplies of dilute sulphuric, hydrochloric and nitric acids were available, which of these acids would you use to remove the scale from the boiler? Give reasons for your choice. (CAM)

8 Rain water changes considerably before it eventually returns to the sea. Describe some of these changes and explain how they occur.

9 Dehydration of a substance is best described as: (a) the removal of moisture; (b) the addition of water; (c) the removal of water; (d) the addition of moisture; (e) the removal of the elements of water.

10 More non-metals

Several important principles, such as the electrochemical series and periodicity, are used in the study of metals and their compounds. Such principles are not quite so important with respect to the non-metals and their compounds. Examination questions on the non-metals tend to depend upon redox reactions, acid-base behaviour and gas preparations, and so these aspects are emphasized in this Unit.

10.1 Carbon and silicon

Carbon – the element

Physical properties Carbon occurs in the pure state as the two allotropic (polymorphic) forms diamond and graphite (p. 95), and in many impure forms such as coal, coke, soot and charcoal. Diamond is the hardest known natural substance, and this is a consequence of its structure.

Graphite also occurs naturally, but most of that now used is made by the Acheson process, in which impure carbon is heated with sand in an electric furnace. Its structure makes a marked contrast to that of diamond; it is a very soft solid and is used as a lubricant, in electrodes, as a moderator in atomic reactors and (mixed with clay) as the 'lead' in pencils.

Powdered forms of carbon have good adsorptive power and are used to purify substances. All forms of the element have giant atomic structures, and thus have high melting points (around 3700 °C).

Chemical properties Carbon atoms need to lose, gain or share 4 electrons in order to achieve a stable electronic structure, and so carbon is not very reactive and always forms covalent bonds.

1 When heated to a high temperature in a plentiful supply of air or oxygen, it glows and produces carbon dioxide,

$$C(s) + O_2(g) \rightarrow CO_2(g)$$

2 When carbon is heated strongly in carbon dioxide, it is reduced to carbon monoxide. This happens in the blast furnace.

$$C(s) + CO_2(g) \rightarrow 2CO(g)$$

3 It will reduce the oxides of unreactive metals to the metal. A high temperature is needed. You may have done this on a small scale using a blowpipe, charcoal block and (e.g.) lead oxide,

$$PbO(s) + C(s) \rightarrow Pb(l) + CO(g)$$

4 When carbon is heated to white heat it reacts with steam,

$$C(s) + H_2O(g) \rightarrow CO(g) + H_2(g)$$

The gases produced are both flammable, and the mixture was traditionally used as an industrial fuel ('water gas') and to manufacture hydrogen.

Carbon dioxide

Preparation Generator B and collection J or H (see Figure, p. 83). Solid reagent: calcium carbonate (limestone or marble chips); liquid reagent: dilute hydrochloric acid.

$$CaCO_3(s) + 2HCl(aq) \rightarrow CaCl_2(aq) + CO_2(g) + H_2O(l)$$

In industry the gas is made by heating limestone in a lime kiln and is also obtained as a byproduct of fermentation reactions.

Physical properties

Solubility in water	Colour	Odour	Density relative to air	Toxicity
Slight. More soluble under pressure – used in 'fizzy' drinks	Colourless	Virtually none	More dense	Only toxic at relatively high concentrations

Chemical properties

1 Normally the gas does not support combustion (e.g. it puts out a lighted taper) but burning magnesium metal continues to burn in it,

$$2Mg(s) + CO_2(g) \rightarrow 2MgO(s) + C(s)$$

A crackling is heard, and white magnesium oxide is formed, together with black specks of carbon. *Note:* The extinguishing of a lighted taper is *not* a test for the gas; other gases do this, e.g. nitrogen. The test for carbon dioxide is given on p. 169.

2 It is weakly acidic, and turns damp blue litmus paper a wine-red colour. Its solution in water *behaves* as if it were carbonic acid,

$$H_2O(l) + CO_2(g) \rightarrow H_2CO_3(aq)$$

but the acid is weak, unstable, and cannot be isolated from solution.

Uses of carbon dioxide

1 It is dissolved in water under pressure to make 'fizzy' drinks.
2 The solid form (dri-ice) is used as a refrigerant.
3 It is used in fire extinguishers; being a dense gas, it blankets the fire and prevents oxygen from reaching it.
4 It is produced 'in situ' by baking powder, and also in health salts. Baking powder consists of a dry mixture of sodium hydrogencarbonate and a solid acid such as tartaric or citric acid. Reaction only takes place when water is added, when the acid reacts with the hydrogencarbonate to form carbon dioxide. A similar principle is used in health salts.

Carbon monoxide, CO
Physical properties

Solubility in water	Colour	Odour	Density relative to air	Toxicity
Insoluble	None	None	Slightly less dense	Toxic; particularly dangerous as has no smell

Chemical properties

1 Burns in air or oxygen with a blue flame to form carbon dioxide,

$$CO(g) + \tfrac{1}{2}O_2(g) \rightarrow CO_2(g)$$

This reaction can be used to distinguish it from carbon dioxide, which does not burn.

2 Good reducing agent. Used in industry to reduce oxides of some metals to the metal, e.g. in the blast furnace. This is the only important use of the gas.
3 Unlike carbon dioxide, it is a neutral oxide and has no reaction with acids, alkalis or indicators.

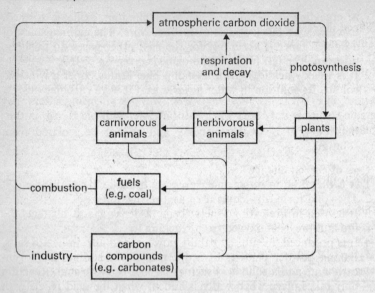

The carbon cycle

The carbon cycle
Carbon is an essential constituent of all living things, and carbon dioxide is the 'transfer agent' for the carbon which is continually circulating in nature. There is a balance between the carbon dioxide liberated into the atmosphere and that used up from the atmosphere. This is summarized in the form of a carbon cycle, such as that shown in the diagram. Remember that virtually all of the carbon atoms on Earth have been present since the Earth began, and that these same atoms are constantly circulating in nature.

Silicon and its compounds
(*not included in all syllabuses*)
Silicon – the element
Like carbon, silicon is in Group 4 of the Periodic Table, has a valency of 4 and is relatively unreactive. The element occurs widely in nature in combined form (e.g. sand, clay). In the pure state it is a hard, brittle solid and is used in the manufacture of silicon chips, silicon steels (chemically resistant steels) and in the preparation of silicones. Some of its compounds are important.

Silicon dioxide, SiO_2

This occurs naturally in several forms such as quartz, flints and sand. It has a giant atomic structure and is thus a solid at room temperature, with a high melting point, contrasting markedly with the molecular substance carbon dioxide. Like carbon dioxide, it is acidic, but as it is insoluble in water it will not affect indicators; its acidic nature is only seen when it forms salts with bases, e.g. in the blast furnace,

$$CaO(s) + SiO_2(s) \rightarrow CaSiO_3(l) \text{ (slag)}$$

and in the manufacture of glass.

Silicon compounds in the manufacture of glass
Ordinary glass (soda glass)

Manufacture and composition	*Additives*	*Advantages or disadvantages*
It is a mixture of sodium silicate and calcium silicate, made by heating a mixture of sand (silicon dioxide), sodium carbonate and calcium carbonate in a furnace at 1400 °C. The molten glass is machined to specifications	Transition metal oxides added to make coloured glass. Aluminium oxide or boron oxide gives glass with a low coefficient of expansion, e.g. Pyrex	Cheap glass is often tinted yellow or green by impurities in the sand. Without additives, it is easily cracked by rapid heating or cooling

Silica glass This is made by melting pure silicon dioxide. It can be heated to 1500 °C before it softens, and has a very low coefficient of expansion, i.e. it withstands rapid heating and cooling. It is more transparent to visible, infra red and ultra violet light, and is used for optical instruments.

10.2 Sulphur and its compounds

Sulphur – the element

Extraction Over 80% of the world supply of sulphur is extracted from deposits of the element in Texas and Louisiana. The sulphur cannot be mined by conventional methods as it occurs below quicksands, and so it is extracted by the Frasch process (see Figure). The tube is about 15 cm in diameter.

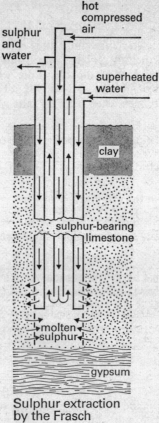

Sulphur extraction by the Frasch process

Physical properties Yellow solid, insoluble in water but soluble in solvents such as carbon disulphide. Its allotropic forms and changes on heating are discussed on p. 98.

Chemical properties The only reaction which you are likely to encounter is its combustion in air or oxygen (p. 84).

Uses Large quantities are used in the manufacture of sulphuric acid (p. 206) and vulcanized rubber; smaller amounts are used to make matches, gunpowder, drugs and sulphur-based insecticides and fungicides.

Sulphur dioxide, SO$_2$
Preparation Generator A and collection H. Liquid reagent: concentrated sulphuric acid; solid reagent: copper turnings.

$$Cu(s) + H_2SO_4(l) \xrightarrow{heat} CuSO_4(aq) + SO_2(g) + 2H_2O(l)$$

Collection in fume cupboard.

Physical properties

Solubility in water	Colour	Odour	Density relative to air	Toxicity
Quite soluble. Gives fountain experiment	None	Unpleasant, sharp smell, leaving metallic taste	More dense	Toxic

Note: The gas is easily liquefied under slight pressure.

Chemical properties
As a reducing agent The gas is a good reducing agent if water is present, when it dissolves to form sulphurous acid,

$$SO_2(g) + H_2O(l) \rightarrow H_2SO_3(aq)$$

which dissociates to produce sulphite ions, $SO_3^{2-}(aq)$. The reductions brought about by sulphur dioxide are thus better thought of as reactions of the sulphite ion, which accepts oxygen from the substance being reduced and is itself oxidized to the sulphate ion,

$$SO_3^{2-}(aq) + [O] \rightarrow SO_4^{2-}(aq)$$

Alternatively, you can consider the reaction in terms of electron transfer if this is more convenient,

$$SO_3^{2-}(aq) + H_2O(l) - 2e \rightarrow SO_4^{2-}(aq) + 2H^+(aq)$$

You have probably seen some of the following examples of these reducing properties. Make sure that you can explain what is being oxidized and reduced in each case.

$$H_2O_2(aq) + SO_3^{2-}(aq) \rightarrow SO_4^{2-}(aq) + H_2O(l)$$

(or, if it is easier to remember as two separate half equations,

$$H_2O_2(aq) \rightarrow H_2O(l) + [O], \text{ and } SO_3^{2-}(aq) + [O] \rightarrow SO_4^{2-}(aq))$$

$$Br_2(aq) + SO_3^{2-}(aq) + H_2O(l) \rightarrow 2Br^-(aq) + SO_4^{2-}(aq) + 2H^+(aq)$$

(in bromine water; brown-yellow) (colourless)

potassium manganate(VII) + SO_3^{2-}(aq) → colourless solution
solution, containing   containing Mn^{2+} ion
purple MnO_4^- ion   + SO_4^{2-}(aq)

potassium dichromate(VI) + SO_3^{2-}(aq) → green solution
solution, containing   containing Cr^{3+} ion
orange $Cr_2O_7^{2-}$ ion   + SO_4^{2-}(aq)

As a bleaching agent Some substances are bleached when oxygen is added to them, i.e. when they are oxidized. This is why chlorine and hydrogen peroxide (both good oxidizing agents) are good bleaching agents. It is also possible to decolourize other substances by removing oxygen, i.e. by reducing them. Moist sulphur dioxide can bleach some wool and silk dyes, straw and flowers by reducing them. It is used to bleach wood pulp for making newspapers. Such reactions are not permanent for the substances are reoxidized slowly in the air, which is why old newspapers go yellow.

As an oxidizing agent Sulphur dioxide is normally a reducing agent, and it acts as an oxidizing agent only if it reacts with a substance which is a more powerful reducing agent than itself. This happens when gas jars of sulphur dioxide and hydrogen sulphide gases are mixed and a little water is introduced,

$$2H_2S(g) + SO_2(g) \rightarrow 2H_2O(l) + 3S(s)$$

Note that some of the sulphur is formed by oxidation of the hydrogen sulphide, and some by reduction of the sulphur dioxide.

Sulphur trioxide, (sulphur (VI) oxide) SO_3
Preparation See Figure; hot vanadium(V) oxide is used as a catalyst.

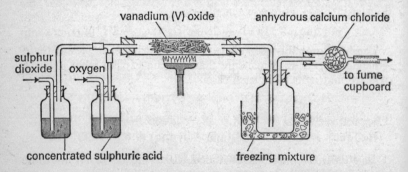

Laboratory preparation of sulphur trioxide (sulphur(VI) oxide) SO_3

The reactant gases must be dried, and atmospheric moisture prevented from reaching the product. The oxide collects as a white solid.

$$2SO_2(g) + O_2(g) \rightarrow 2SO_3(s)$$

Physical properties The white solid melts at 17 °C and boils at 44 °C, but because it fumes strongly with moist air it gives the appearance of being a gas at room temperature.

Chemical properties It is an acidic oxide, reacting violently with water to form sulphuric acid.

$$SO_3(s) + H_2O(l) \rightarrow H_2SO_4(aq)$$

This reaction is so violent that a mist of acid droplets forms rather than a solution, and so when sulphuric acid is prepared or manufactured the sulphur trioxide is dissolved in fairly concentrated sulphuric acid rather than water. Absorption then takes place more efficiently, and a very concentrated acid is produced.

Dilute sulphuric acid, $H_2SO_4(aq)$
This acid is a typical acid and its reactions have been covered extensively in the appropriate units.

Concentrated sulphuric acid, $H_2SO_4(l)$
Preparation and manufacture Sulphur trioxide is dissolved in fairly concentrated sulphuric acid solution, so that the solution becomes more concentrated.

Physical properties The pure acid is a colourless, viscous liquid; the concentrated acid contains about 98% of the pure liquid. A great amount of heat is evolved when the concentrated acid comes into contact with water, and when the acid is being diluted it *must* be added to the water. If water is added to the acid, some water is quickly vaporized to steam and 'spits out' with corrosive droplets of the acid.

Chemical properties
Note: These reactions are not given by the dilute acid.

1 Its affinity for water makes it ideal for drying gases, but not those which react with it, such as ammonia.
2 The concentrated acid will remove chemically combined water,

i.e. it is a dehydrating agent, e.g. when added to hydrated copper(II) sulphate crystals or crystals of a sugar.

$$CuSO_4 \cdot 5H_2O(s) \xrightarrow{-5H_2O} CuSO_4(s) + steam$$
(blue crystals) (white powder)

$$C_6H_{12}O_6(s) \xrightarrow{-6H_2O} 6C(s) + steam$$
(white crystals) (black solid)

3 The concentrated acid is a powerful oxidizing agent, especially when hot, and is itself reduced to sulphur dioxide in the process.

Uses of sulphuric acid The acid is one of the most important industrial chemicals, and is used in the manufacture of rayon, fertilizers, dyes, plastics, drugs and explosives. When extra sulphur trioxide is dissolved in it, a very corrosive, fuming liquid called oleum is produced, which is used in the organic chemicals industry.

Sulphurous acid (H_2SO_3) and the sulphites

Sulphurous acid is a weak, unstable acid which cannot be isolated as a pure substance. It is only encountered in aqueous solution. Its solution, like those of the sulphites, has the same reactions as moist sulphur dioxide.

10.3 Nitrogen and phosphorus

Nitrogen – the element

Preparation Impure nitrogen can be prepared in the laboratory from the air (p. 78) and industrially from liquid air (p. 210). The pure gas is prepared using generator C and collection J, for which concentrated solutions of ammonium chloride and sodium nitrite are mixed and warmed. The ammonium nitrite formed then decomposes,

$$NH_4NO_2(aq) \rightarrow N_2(g) + 2H_2O(l)$$

Physical properties

Solubility in water	Colour	Odour	Density relative to air	Toxicity
Virtually insoluble	None	None	Same	Non-toxic

Chemical properties The chief chemical characteristic of nitrogen is its inertness. The only reaction you are likely to study is its combination with hydrogen in the Haber process (p. 205). There is no positive test for the gas.

Uses of nitrogen
1 It is converted by the Haber process into ammonia, and subsequently into fertilizers and nitric acid.
2 Liquid nitrogen is used as a refrigerant.
3 It provides an inert atmosphere for certain chemical reactions.

Oxides of nitrogen
Nitrogen forms several oxides, but the only one you are likely to encounter in the laboratory is the dioxide, NO_2. (Nitrogen monoxide, NO, occurs as an intermediate in the industrial manufacture of nitric acid.)

The preparation of nitrogen dioxide (dinitrogen tetroxide) The gas is frequently observed as a product of the action of heat on metallic nitrates and the reaction of nitric acid on metals. The generator is a boiling tube containing solid lead nitrate, connected to collection system G in a fume cupboard.

$$2Pb(NO_3)_2(s) \xrightarrow{heat} 2PbO(s) + 4NO_2(g) + O_2(g)$$

The nitrogen dioxide liquefies and collects in the U-tube, and the oxygen escapes. The gas is then obtained by warming the liquid gently.

Physical properties

Solubility in water	Colour	Odour	Density relative to air	Toxicity
Reacts and dissolves to form acidic solution	Dark brown gas or pale yellow-brown liquid	Sharp, unpleasant	More dense	Very toxic

Chemical properties
1 It reacts with water to form a mixture of nitric and nitrous acids.

$$2NO_2(g) + H_2O(l) \rightarrow HNO_2(aq) + HNO_3(aq)$$

2 The pale yellow-brown liquid form (N_2O_4) dissociates when heated into the dark brown gas, NO_2. This reaction is sometimes studied in connection with equilibria.

$$N_2O_4(l) \rightleftharpoons 2NO_2(g)$$

Ammonia, NH_3

Preparation Generator C + drying system F + collection I. Solid reagents: a mixture of dry calcium hydroxide and dry ammonium chloride. The drying agent must be calcium oxide.

$$Ca(OH)_2(s) + 2NH_4Cl(s) \rightarrow CaCl_2(s) + 2NH_3(g) + H_2O(l)$$

Physical properties

Solubility in water	Colour	Odour	Density relative to air	Toxicity
Extremely soluble	None	Unpleasant, choking	Less dense	Toxic

Note:
1 The extreme solubility of ammonia is normally shown by the 'fountain experiment'. You could be asked to describe such an experiment in detail, and to explain clearly how the fountain is created.
2 Ammonia is easily liquefied under pressure.

Chemical properties
1 Ammonia is the only common alkaline gas, and this property is used as a test (see p. 169). It neutralizes acid solutions and acid gases to form ammonium salts, e.g.

$$NH_3(g) + HCl(g) \rightarrow NH_4Cl(s)$$

2 It will reduce the heated oxides of metals below iron in the activity series to the metal, e.g.

$$2NH_3(g) + 3CuO(s) \rightarrow 3Cu(s) + 3H_2O(g) + N_2(g)$$

Uses of ammonia Most of the ammonia manufactured is converted into nitric acid (p. 210) and into fertilizers. It is also used to prepare urea (for fertilizers and plastics) and, when liquefied, as a refrigerant.

Ammonia solution, $NH_3(aq)$, and ammonium salts

Preparation of ammonia solution Ammonia gas is prepared in the normal way and, instead of being dried and collected, is passed into water as shown in the Figure. A solution of any other very soluble gas (e.g. hydrogen chloride dissolving to produce hydrochloric acid) is made in the same way. If the gas was passed straight into water, 'sucking back' could occur. You could be asked to explain how sucking back is prevented by using the inverted funnel.

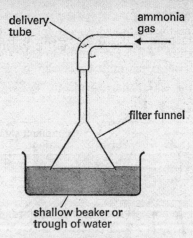

Preparation of a solution of ammonia

Properties of ammonia solution and ammonium salts
1 Ammonia solution is an important (although weak) alkali. It shows the usual properties of alkalis, neutralizing acids to form ammonium salts and precipitating metal hydroxides from solutions of their salts (but not without complications, as explained on p. 117).
2 All ammonium salts are soluble in water, liberate ammonia when heated with an alkali (used as a test for an ammonium compound and also to prepare ammonia) and undergo either thermal dissociation or thermal decomposition when heated, e.g.

$NH_4Cl(s) \rightleftharpoons NH_3(g) + HCl(g)$ (thermal dissociation)

$(NH_4)_2CO_3(s) \rightarrow 2NH_3(g) + CO_2(g) + H_2O(g)$
(thermal decomposition)

Uses of ammonium compounds
1 Ammonium chloride, NH_4Cl, is used in dry batteries (electrolyte) and in soldering.
2 Ammonia solution is used in household cleaning agents.
3 Ammonium carbonate is used in smelling salts.
4 Ammonium nitrate is used in large quantities as a fertilizer, and also in explosives.
5 The sulphate is widely used as a fertilizer.

Nitric acid, HNO_3

Laboratory preparation Although this is not included in some syllabuses, a typical preparation is shown in the Figure. Excess heating

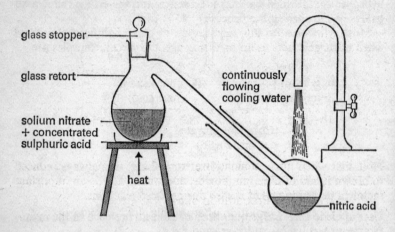

Laboratory preparation of nitric acid

is avoided, to prevent thermal decomposition of the acid. An all-glass apparatus must be used as the concentrated acid rapidly attacks cork and rubber.

$$H_2SO_4(l) + KNO_3(s) \rightarrow HNO_3(g) + KHSO_4(s)$$

Physical properties The acid prepared as above contains more than 90% nitric acid, and is called fuming nitric acid. Concentrated nitric acid contains about 68% nitric acid. The pure acid is a dense, oily, colourless liquid, but it and its concentrated solution are often yellow because of the presence of oxides of nitrogen formed by the slow decomposition of the acid in daylight. This is why the concentrated acid is often stored in brown bottles.

The chemical properties of dilute nitric acid The dilute acid is a typical strong acid except in its reactions with metals.

The chemical properties of concentrated nitric acid
Thermal decomposition When heated, the concentrated acid decomposes rapidly, forming brown fumes of nitrogen dioxide.

$$4HNO_3(l) \rightarrow 4NO_2(g) + 2H_2O(g) + O_2(g)$$

Oxidizing properties The concentrated acid is a powerful oxidizing agent. (The dilute acid also shows oxidizing properties, which is why it does not liberate hydrogen when added to metals; the hydrogen first formed is oxidized to water, and the acid is reduced to oxides of nitrogen in the process.)

Most syllabuses do not require you to show balanced equations when nitric acid acts as an oxidizing agent. Typical examples are:

$$2I^-(aq) \quad - \quad 2e \quad \to I_2(s)$$
(e.g. from potassium iodide solution) (to concentrated nitric acid)

$$H_2S(g) + \quad [O] \quad \to H_2O(l) + S(s)$$
(from concentrated nitric acid)

Note that when nitric acid oxidizes something, it is always reduced to brown fumes of nitrogen dioxide, and water. This is an important factor in describing what is *seen* during these reactions.

Uses of nitric acid Large quantities of the acid are used in the manufacture of fertilizers, explosives and dyes.

The nitrogen cycle

Nitrogen is an essential constituent of all living organisms. As with the carbon cycle, the world contains a virtually constant number of

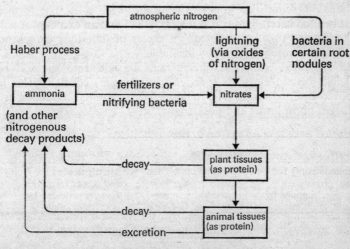

The nitrogen cycle

nitrogen atoms, which are circulating continually in different compounds. Although the air contains 80% nitrogen by volume, this has to be converted into a soluble form (i.e. it has to be 'fixed') before it can be used by most plants. The natural cycle has been disturbed by man's need to produce ever-increasing supplies of food. In constantly removing vegetables, etc. from plots of land, the soil becomes low in nitrogen compounds, and so nitrogen has to be fixed artificially (usually via Haber process → ammonia → fertilizers) and fed to the soil. A typical nitrogen cycle is shown in the diagram.

Phosphorus – the element

Phosphorus exists as a solid in two main allotropic forms, commonly called white phosphorus and red phosphorus. The element is not studied in detail at an elementary level, although some syllabuses include its reactions with oxygen and chlorine.

Fertilizers. NPK values. Soil pH

A wide variety of chemicals are available for use in gardening, and some of these are very simple, e.g. fertilizers and those chemicals used to control the pH of soil.

Although an acid soil is useful for growing certain plants, it is undesirable for growing vegetables, so farmers and gardeners add either calcium hydroxide ('hydrated lime') or powdered calcium carbonate (ground limestone) to neutralize such soils. The former must be used carefully, for it is a weak alkali and an excess of it will make the soil alkaline, whereas an excess of limestone causes no problems because it is virtually insoluble.

As explained earlier, fertilizers must be added to the soil to replenish certain elements (particularly nitrogen) constantly being removed by food crops. The main elements required by plants are nitrogen, phosphorus and potassium.

Nitrogen	*Phosphorus*	*Potassium*
Required mainly for 'green' growth, e.g. cabbages. Supplied by nitrates, ammonium salts and manure	Required mainly for healthy root growth. Provided by 'superphosphate' (a type of calcium phosphate), bone meal, hoof and horn and manure	Mainly needed for fruit and flower formation. Usually provided by potassium sulphate

A balanced fertilizer (e.g. National Growmore) contains all three main elements in roughly equal proportions. Other fertilizers are formulated with different proportions of the elements, according to the particular needs of the crop and the soil. The composition of a fertilizer is often given by using 'NPK' values (for nitrogen, phosphorus and potassium), usually expressed as a percentage by weight of the mixture.

10.4 Chlorine and the halogens

The halogen family
Some of the chemistry of the halogens is described on p. 94. They also undergo displacement reactions with each other, i.e.

$$Cl_2(g) + 2Br^-(aq) \rightarrow 2Cl^-(aq) + Br_2(aq)$$
(colourless, from any bromide) (brown-orange)

$$Cl_2(g) + 2I^-(aq) \rightarrow 2Cl^-(aq) + I_2(s)$$
(colourless, from any iodide) (dark brown)

$$Br_2(l) + 2I^-(aq) \rightarrow 2Br^-(aq) + I_2(s)$$
(brown-orange) (colourless) (colourless) (dark brown)

This type of reaction is similar to the displacement reactions considered with the activity series of metals on p. 59. In this case, a more reactive halogen will displace a less reactive halogen from a solution of one of its salts by 'grabbing' electrons, whereas a metal will displace a less reactive metal from a solution of one of its salts by giving electrons. For example, chlorine will displace bromine from a bromide, but bromine cannot displace chlorine from a chloride.

Chlorine – the element
Preparation Generator B + collection H, in the fume cupboard. Solid reagent: potassium manganate(VII); liquid reagent: concentrated hydrochloric acid.

$$2HCl(aq) + [O] \text{ (from oxidizing agent)} \rightarrow H_2O(l) + Cl_2(g)$$

Physical properties

Solubility in water	Colour	Odour	Density relative to air	Toxicity
Slightly soluble	Green-yellow	Choking, characteristic	More dense	Very toxic

Chemical properties

Direct combination with other elements Chlorine is a very reactive element, and readily combines with many other elements, both metals and non-metals. Two common examples are:

1 *Sodium* If a small piece of sodium is ignited on a combustion spoon, then plunged into a gas jar of chlorine, the metal burns with a yellow flame, forming dense, white clouds of solid sodium chloride,

$$2Na(s) + Cl_2(g) \to 2NaCl(s)$$

2 *Magnesium* If a piece of magnesium ribbon is ignited on the end of a combustion spoon and then placed in a gas jar of chlorine, the metal burns with a white flame, forming a dense white cloud of solid magnesium chloride,

$$Mg(s) + Cl_2(g) \to MgCl_2(s)$$

The affinity of chlorine for hydrogen The gas has a great affinity for hydrogen, and in addition to reacting explosively with the element in sunlight, it reacts with many compounds containing hydrogen. It is easy to work out what happens in such cases, because the chlorine always removes the hydrogen to form hydrogen chloride, depositing or liberating 'whatever is left' from the compound. Typical experiments you may have seen include the following.

	Hydrocarbons		*Hydrogen sulphide*
a wax taper		*turpentine*	
Taper ignited and placed in gas jar of chlorine. Continues to burn with a smoky flame, depositing large amounts of black carbon and forming steamy fumes of hydrogen chloride		Piece of filter paper soaked in warm turpentine and dropped into gas jar of chlorine. The turpentine bursts into a sheet of red flame, forming large amounts of black carbon and steamy fumes of hydrogen chloride $C_{10}H_{16}(l) + 8Cl_2(g) \to 10C(s) + 16HCl(g)$	Gas jars of the two gases are placed mouth to mouth. Instant reaction produces steamy fumes of hydrogen chloride and yellow deposits of sulphur form on the sides of the jars $H_2S(g) + Cl_2(g) \to 2HCl(g) + S(s)$

As an oxidizing agent Chlorine is a powerful oxidizing agent. Chlorine atoms readily gain electrons (i.e. oxidize something else by removing electrons from it) to form stable Cl^- ions, as in the halogen displacement reactions (p. 161) and the conversion of iron(II) salts to iron(III) salts (p. 129), all of which are examples of oxidation. Similarly, its affinity for hydrogen means that it removes hydrogen from compounds, and so the reactions just considered above are also oxidations.

Chlorine as a bleaching agent In the presence of moisture, the gas rapidly bleaches coloured flowers, litmus paper, inks and other materials dyed with vegetable dyes. The chlorine reacts with the moisture to form a mixture of two acids:

$$Cl_2(g) + H_2O(l) \rightarrow HOCl(aq) + HCl(aq)$$

The former, chloric(I) acid (commonly called hypochlorous acid), bleaches substances by oxidizing the colourings in them to a colourless form,

$$HOCl(aq) \rightarrow HCl(aq) + [O]$$

This acidity of chlorine solution, followed by a bleaching action, is used as a test for the gas (p. 169).

The reactions of chlorine with alkalis Chlorine reacts with alkalis in two different ways.

With *cold, dilute* alkali,

$$Cl_2(g) + 2NaOH(aq) \rightarrow NaCl(aq) + NaOCl(aq) + H_2O(l)$$

NaOCl, sodium chlorate(I), is commonly called sodium hypochlorite; salts of this kind act like hypochlorous acid, i.e. they are used in preparing bleaches.

With *hot, concentrated* alkali,

$$Cl_2(g) + 6NaOH(aq) \rightarrow$$
$$5NaCl(aq) + \underset{\text{sodium chlorate(V)}}{NaClO_3(aq)} + 3H_2O(l)$$

Uses of chlorine Large quantities are used in chlorinating organic compounds to make rubbers, plastics (especially PVC) and solvents. It is also used for sterilizing drinking water and sewerage, in preparing dry cleaning fluids and in making hydrochloric acid.

Hydrogen chloride, HCl(g)
Preparation Generator A + collection H, in fume cupboard. Solid reagent: sodium chloride; liquid reagent: concentrated sulphuric

acid. (Warming may be needed after the initial vigorous reaction has subsided.)

$$NaCl(s) + H_2SO_4(l) \rightarrow NaHSO_4(aq) + HCl(g)$$

Physical properties

Solubility in water	Colour	Odour	Density relative to air	Toxicity
Extremely soluble – gives fountain experiment	None, but fumes in air	Sharp and characteristic	More dense	Toxic

Chemical properties These have already been described elsewhere in the book, e.g. its reaction with ammonia (p. 169) is also used as a test.

Note: A solution of hydrogen chloride in a dry, organic solvent (e.g. methylbenzene, as on p. 67) contains hydrogen chloride *molecules* and is not acidic, whereas a solution of hydrogen chloride in water contains $Cl^-(aq)$ and $H^+(aq)$ ions and is acidic.

Hydrochloric acid, HCl(aq)

Preparation Hydrogen chloride gas is prepared as just described and then passed into water via an inverted funnel, as shown for ammonia in the diagram on p. 157.

Properties Hydrochloric acid is a typical strong acid, and its reactions have been described throughout the book. The concentrated acid behaves as a more reactive form of the dilute acid; there are no additional properties like the oxidizing and dehydrating properties of concentrated sulphuric acid, or the oxidizing properties of concentrated nitric acid.

Questions

1 Give one example in each case of sulphur dioxide acting as: (*a*) an oxidizing agent; (*b*) a reducing agent; (*c*) an acid anhydride. (JMB)

2 When sulphur is burned in a porcelain boat in an atmosphere of oxygen, the mass of the boat and contents decreases. On the other hand, when magnesium is burned in the same way, there is an increase in the mass of the boat and contents.
   (*a*) What would you see in the case of sulphur?
   (*b*) Name the major product formed when the sulphur is burned in oxygen.
   (*c*) What would you see in the case of magnesium?
   (*d*) Name the product formed when magnesium is burned in oxygen.

(e) Why does one weighing show a decrease and the other an increase in mass?

(f) Which product might be a liquid at a temperature of 253 K (−20 °C)?   (JMB)

3 Which of the following compounds would make the best nitrogenous fertilizer? (H = 1; N = 14; O = 16; Na = 23; Mg = 24; S = 32; Cl = 35.5).

(a) NH_4Cl; (b) $(NH_4)_2SO_4$; (c) $NaNO_3$; (d) NH_4NO_3; (e) $Mg(NO_3)_2$.

4 Sulphur: (a) forms two alkaline oxides; (b) is spontaneously flammable; (c) burns with a blue flame; (d) conducts electricity in the molten state; (e) is usually stored in the form of sticks in water.

5 Ammonium sulphate is widely used as a fertilizer because: (a) it provides nitrogen for the plants; (b) it provides sulphur for the plants; (c) it provides oxygen for the plants; (d) it poisons harmful bacteria; (e) it helps to break up the soil.

6 Name *one* acidic and *one* alkaline gas which, in each case, is very soluble in water. Draw a diagram to show how a very soluble gas can be safely dissolved in water contained in a beaker and explain briefly how the apparatus avoids the danger of 'sucking back'.   (JMB)

7 Ammonia can be prepared by heating an ammonium salt with an alkali.

(a) Name a pair of reagents suitable for this reaction.

(b) Give the simplest ionic equation for the reaction.

Ammonia is (i) very soluble in water and (ii) less dense than air. How does each of these properties determine the way in which ammonia is collected in a gas jar?   (JMB)

8 (a) Draw a labelled diagram of the apparatus you would use to prepare and collect a sample of hydrogen chloride, starting from sodium chloride. Write an equation for the reaction and state the reaction conditions.

(b) Explain why solutions of ammonia and hydrogen chloride in water are alkaline and acidic respectively whereas dry ammonia and dry hydrogen chloride have no effect on dry red or blue litmus paper.

(c) Describe and explain the changes that take place when a little ammonium chloride is heated in a test-tube.

9 Describe and illustrate the laboratory preparation and collection of sulphur dioxide from sulphuric acid. Explain how you would convert sulphur dioxide into sulphur trioxide. How do the two oxides differ physically?

Give a chemical test to distinguish between the solutions obtained when each of these oxides is dissolved in water.

The result of the test should be given for each solution.   (JMB)

11 Analysis

11.1 Volumetric analysis

Calculations on molarities are discussed on p. 102, and an understanding of these is essential for this work.

Preparing a standard solution

Suppose that it is required to make up 250 cm^3 of a solution containing 0.1 mol dm^{-3} (0.1 M) of anhydrous sodium carbonate.

1 Work out the mass of anhydrous sodium carbonate needed to make up the solution.
   The mass of 1 mole of anhydrous sodium carbonate,

   $$Na_2CO_3 = (2 \times 23) + 12 + (3 \times 16) = 106 \text{ g}.$$

   (This would be 286 g if hydrated sodium carbonate, $Na_2CO_3.10H_2O$, were used.)
   ∴ If 106 g of sodium carbonate (anhydrous salt) are dissolved and made up to 1 dm^3 of solution, the solution would have a concentration of 1 mol dm^{-3} (1 M) of sodium carbonate.
   ∴ If 10.6 g are dissolved and made up to 1 dm^3 of solution, the solution would have a concentration of 0.1 mol dm^{-3} (0.1 M).
   ∴ 250 cm^3 of a solution having the same concentration would need $\frac{1}{4} \times 10.6$ g = 2.65 g.

2 Weigh out accurately, on a watch glass, a mass of anhydrous sodium carbonate *approximately* equal to the calculated mass.
3 Dissolve *all* of this material in deionized water, in a beaker, and transfer *all* of the solution into a 250 cm^3 volumetric flask. Carefully add deionized water until the bottom of the meniscus is exactly on the graduation mark. Mix thoroughly.
   Note: If it is relevant to your syllabus, you should be able to describe in detail the procedures you would use to ensure that all

of the original solid ends up in the final solution, e.g. rinsing the stirring rod, beaker, etc.
4 Calculate the *actual* concentration which your final solution will have.

Performing a titration

In volumetric analysis, one solution of known concentration is reacted with a second solution of unknown molarity in order to determine the concentration of the latter. A pipette and burette are used, and the solutions are reacted together under controlled conditions in order to find the reacting volumes.

Some syllabuses could require you to describe in detail the various steps used in performing a titration. The following points are sometimes forgotten or misunderstood.

1 All readings must be taken at eye level, and with the apparatus vertical.
2 It is not necessary to dry the conical flask between operations.
3 It is always advisable to perform a rough titration first.
4 Shake the flask during addition of the liquid from the burette.
5 Repeat operations until two or more readings agree within 0.1 cm^3.
6 When using a strong acid (sulphuric, hydrochloric or nitric) with a strong alkali (sodium or potassium hydroxides) any indicator is suitable. When titrating a strong acid with sodium or potassium carbonate solution, methyl orange must be used as indicator.
7 This technique can also be used to prepare some soluble salts, as described on p. 73.

Calculations using titration results

1 Ensure that you know the concentration of *one* of the two solutions involved. If this is not given directly, it will be necessary to work it out from the mass of solid dissolved in a certain volume of solution.
2 Suppose that 25.0 cm^3 of a solution of hydrochloric acid of concentration 0.1 mol dm^{-3} (0.1 M) react with 21.5 cm^3 of a solution of sodium hydroxide, and we need to calculate the concentration of the sodium hydroxide solution.
3 Write down the balanced equation for the reaction:

$$HCl(aq) + NaOH(aq) \rightarrow NaCl(aq) + H_2O(l)$$

4 Calculate the number of moles used in the titration for the chemical the concentration of which is known, e.g.

25.0 cm³ of hydrochloric acid used, of concentration 0.1 mol dm⁻³ (0.1 M).
1000 cm³ of the hydrochloric acid solution contain 0.1 moles.
∴ 1 cm³ of the hydrochloric acid solution contains

$$\frac{1}{1000} \times 0.1 \text{ moles.}$$

∴ 25 cm³ of the acid solution contains $25 \times \frac{1}{1000} \times 0.1$ moles.

5 From the equation, calculate how many moles of the *other* substance this has reacted with. In this example, one mole of hydrochloric acid reacts with one mole of sodium hydroxide.

∴ If $\frac{25 \times 0.1}{1000}$ moles of acid are used, $\frac{25 \times 0.1}{1000}$ moles of alkali will react with it.

6 Calculate how many moles of this second substance would be contained in 1000 cm³ of its solution; this is the concentration required, e.g.

$\frac{25 \times 0.1}{1000}$ moles of alkali are used. This was in 21.5 cm³ of solution.

∴ 1 cm³ of the alkali solution contains $\frac{25 \times 0.1}{1000 \times 21.5}$ moles.

∴ 1000 cm³ of the alkali solution contains

$$\frac{1000 \times 25 \times 0.1}{1000 \times 21.5} \text{ moles} = \frac{25 \times 0.1}{21.5} \text{ moles} = 0.11 \text{ moles.}$$

The alkali solution has a concentration of 0.11 mol dm⁻³ (0.11 M).
Note: It is absolutely vital to base your calculation on a correctly balanced equation. If the same experimental results had been obtained for a reaction between hydrochloric acid and sodium carbonate solution, the final answer (i.e. the concentration of the sodium carbonate solution) would be different, because the equation for this reaction shows that the reagents react in the ratio 1:2 instead of 1:1 as in the first example.

$$2HCl(aq) + Na_2CO_3(aq) \rightarrow 2NaCl(aq) + CO_2(g) + H_2O(l)$$

In step (4), therefore, the number of moles of sodium carbonate reacting would be *half* the number of moles of acid used, i.e.

Analysis   169

$\dfrac{1 \times 25 \times 0.1}{2 \times 1000}$ moles, and this would be used in the final steps rather than $\dfrac{25 \times 0.1}{1000}$.

11.2   Tests for common gases and ions

Some of these have been included in other parts of the book, but it is also convenient to include them altogether here (Tables 11.1–11.4) as it is quite common for an examination question to be based entirely on tests of this type.

Table 11.1   *Tests for some common gases*

Gas	Test	Result of test if positive
Hydrogen	Trap gas in test-tube, apply lighted taper	Squeaky pop
Oxygen	Apply glowing taper to gas sample in test-tube	Taper relights
Carbon dioxide	Pass gas into calcium hydroxide solution (lime water), e.g. by collecting sample in teat pipette and ejecting into small volume of calcium hydroxide solution in test-tube	Calcium hydroxide solution goes 'milky' due to fine precipitate of calcium carbonate
Ammonia	1 Expose gas to damp red litmus paper, or 2 Expose gas to fumes of hydrogen chloride, e.g. from a bottle of concentrated hydrochloric acid	1 Paper goes blue 2 Dense white cloud of ammonium chloride formed
Chlorine	Expose gas to damp blue litmus paper	Paper goes pink and is then bleached
Hydrogen chloride	Expose gas to ammonia fumes, e.g. from bottle of concentrated ammonia solution	Dense white cloud of ammonium chloride
Nitrogen	Apply all other gas tests	If none positive, gas probably nitrogen
Sulphur dioxide	Expose gas to filter paper soaked in acidified potassium dichromate(VI) solution	Paper changes from orange to green
Hydrogen sulphide	Expose gas to filter paper soaked in lead nitrate (or acetate) solution	Dark coloured stain of lead sulphide forms on paper

Table 11.2 *Tests for some common anions*

Anion	Test	Result of test if positive
Soluble chloride	Acidify solution with dilute nitric acid and then add silver nitrate solution	White precipitate of silver chloride formed
Soluble sulphate	Acidify solution with dilute hydrochloric acid and then add barium chloride solution	White precipitate of barium sulphate formed
Nitrates	Make solution alkaline with dilute sodium hydroxide solution, add Devarda's alloy, and warm. (See also brown ring test, p. 121)	Ammonia gas formed (test in usual way)
Carbonates and hydrogen-carbonates	Add a little dilute hydrochloric acid	Effervescence, carbon dioxide gas given off (test in usual way)

Table 11.3 *Some flame tests for metal ions*

Test	Result if positive
Clean nichrome or platinum wire by repeated dipping in hydrochloric acid and roaring bunsen flame. When no colour given to flame by wire, moisten the wire with dilute hydrochloric acid and pick up small sample of compound on it. Hold wire and sample in colourless flame	Intense golden yellow: Na^+ Apple green: Ba^{2+} Green-blue: Cu^{2+} Brick red: Ca^{2+} Lilac: K^+

Table 11.4 *Other tests for positive ions in solution*

Note: For an explanation of the reactions involved in the tests for metal ions, see Table 8.3, p. 117.

Ion	Test	Result if positive
Ammonium	Add dilute sodium hydroxide solution, then warm	Ammonia gas produced (test in usual way)
Copper(II)	Add dilute ammonia solution, dropwise, with stirring	Pale blue precipitate of copper(II) hydroxide formed initially, but then dissolves to form deep blue solution
Zinc	Add dilute ammonia solution, dropwise, with stirring to one sample and add dilute sodium hydroxide in a similar way to a second sample	In *both* samples, a white precipitate (zinc hydroxide) forms initially but then dissolves in excess alkali

Table 11.4 – Contd.

Ion	Test	Result if positive
Iron(II)	Add dilute ammonia or dilute sodium hydroxide solution	Dirty-green precipitate of iron(II) hydroxide
Iron(III)	As with iron(II)	Red-brown precipitate of iron(III) hydroxide
Aluminium and lead(II)	1 Do test as for zinc 2 If necessary (see results column) add dilute hydrochloric acid to a third sample	If white precipitate forms in both cases but only dissolves in excess sodium hydroxide, solution could contain Pb^{2+} or Al^{3+}. Do test (2), no precipitate if Al^{3+}, white precipitate (of lead chloride) if Pb^{2+} present

Questions

1 Which of the following gives a white precipitate with barium chloride solution, and a brick-red flame test? (a) sodium sulphate; (b) copper sulphate; (c) calcium sulphate; (d) ammonium sulphate; (e) calcium chloride.

2 Samples are taken from bottles labelled washing soda, calcium chloride, potassium nitrate, iron(II) sulphate, calcium sulphate, manganese(IV) oxide (manganese dioxide), lead(II) oxide (lead monoxide), iron(III) oxide.
 Identify five of these samples from the following descriptions: (a) green crystals; (b) damp white lumps; (c) a black powder; (d) large colourless crystals covered in white powder; (e) a red-brown powder. (JMB)

3 Bottles containing three white powders, sodium carbonate, calcium hydroxide and ammonium chloride respectively, have lost their labels. Using only water and dilute hydrochloric acid as additional reagents, what experiments would you perform in order to re-label the bottles correctly? (JMB)

4 In a reaction between a solution of a metallic hydroxide (formula MOH) and dilute hydrochloric acid, 20.0 cm³ of the alkali reacted with 25.0 cm³ of the acid. The acid concentration was 4.00 g per litre and that of the alkali was 7.67 g per litre. Calculate the formula mass of the alkali and hence the atomic mass of the metal.
 Describe in detail how you would determine the volume of the acid and alkali which exactly neutralize each other. (JMB)

5 When exploded with excess oxygen, ethane (C_2H_6) reacts as in the equation:
$$2C_2H_6(g) + 7O_2(g) \rightarrow 4CO_2(g) + 6H_2O(l)$$
If 20 cm³ of ethane are exploded with 100 cm³ of oxygen and the gaseous products are reduced to the original laboratory temperature and pressure, what will be the volume and composition of the resulting mixture of gases? (CAM)

6 An experiment showed that when 0.1 formula mass of sodium hydrogencarbonate was heated to constant mass, 5.3 g of solid remained. Show that these results agree with the following equation:
$$2NaHCO_3(s) \rightarrow Na_2CO_3(s) + H_2O(g) + CO_2(g)$$
What fraction of a mole of carbon dioxide molecules should have been given off in the above experiment?
 What volume would the carbon dioxide occupy at s.t.p? (WEL)

12 Organic chemistry

Organic chemistry is the study of the chemical compounds of carbon, excluding compounds such as the oxides of carbon and the carbonates.

12.1 Hydrocarbons

The homologous series of alkanes
*An **homologous** series is a group of organic chemicals, all of which have the same general formula and similar chemical properties.*

The alkanes are a series of saturated hydrocarbons, of general formula C_nH_{2n+2}. (A **saturated compound** contains only single covalent bonds.) The early members are gases at room temperature and pressure, but as the molecular mass increases the members become liquids and eventually solids. Some typical alkanes are shown below. Methane is normally studied as a typical member of the series.

	Methane	Ethane	Propane	Butane	Pentane	Hexane	Octane
Formula	CH_4	C_2H_6	C_3H_8	C_4H_{10}	C_5H_{12}	C_6H_{14}	C_8H_{18}
State (room temp. and pressure)	gas	gas	gas	gas	liquid	liquid	liquid

The alkanes provide simple examples of **isomerism**. *Isomers are substances which have the same molecular formula but a different arrangement of atoms within their molecules.*

Organic chemistry 173

For example, there are two isomers of butane, each of which has the molecular formula C_4H_{10}:

```
    H  H  H  H
    |  |  |  |
H — C— C— C— C — H    butane, and
    |  |  |  |
    H  H  H  H


    H  H  H
    |  |  |
H — C— C— C — H     2-methylpropane (isobutane).
    |  |  |
    H  |  H
      H—C—H
        |
        H
```

Make sure that you understand the difference between *isomers*, *isotopes*, and *allotropes*.

Physical properties of methane, CH_4

Solubility in water	Colour	Odour	Density relative to air	Toxicity
Insoluble	None	None	Less dense	Not toxic

Chemical properties
Substitution reactions with halogen elements Alkanes are saturated compounds, and saturated compounds usually take part in substitution reactions.

A substitution reaction is when an organic molecule reacts with an element or compound of the type X—Y so that X enters the organic molecule, replacing an atom (usually hydrogen) which then combines with Y.

Methane and the other alkanes readily take part in substitution reactions with chlorine and bromine. Bromine reacts slowly in diffused daylight, and chlorine more quickly; the reactions can be explosive in sunlight. In each case a series of steps occurs as a hydrogen atom is substituted by a halogen atom, e.g.

$$CH_4(g) + Cl_2(g) \rightarrow \underset{\text{monochloromethane}}{CH_3Cl(l)} + HCl(g)$$

and then $CH_3Cl(l) + Cl_2(g) \rightarrow CH_2Cl_2(l) + HCl(g)$
dichloromethane

and then $CH_2Cl_2(l) + Cl_2(g) \rightarrow CHCl_3(l) + HCl(g)$
trichloromethane

and then $CHCl_3(l) + Cl_2(g) \rightarrow CCl_4(l) + HCl(g)$
tetrachloromethane

Similar reactions occur with bromine, and if other alkanes are used the principle is the same but the number of steps is increased.

Note: It is often useful to show a graphical formula in an equation rather than a molecular formula, e.g.

$$H-\underset{\underset{H}{|}}{\overset{\overset{H}{|}}{C}}-H$$

rather than CH_4, for it is easier to see what is happening to individual bonds during the reaction.

Combustion In a plentiful supply of oxygen, methane burns completely with a clean blue flame to form carbon dioxide and water vapour. Other alkanes react in exactly the same way, e.g

$$CH_4(g) + 2O_2(g) \rightarrow CO_2(g) + 2H_2O(g)$$
$$C_4H_{10}(g) + 6\tfrac{1}{2}O_2(g) \rightarrow 4CO_2(g) + 5H_2O(g)$$

Uses The combustion reactions of alkanes are very exothermic, and this is why many of the alkanes are used as fuels, e.g. natural gas (methane), petrol (a mixture of several alkanes), paraffin (another mixture, of less volatile alkanes), calor gas and camping gas (mainly butane, liquefied under pressure), etc.

Higher alkanes are also used as solvents and in the manufacture of other chemicals. Solid alkanes have a variety of uses, e.g. vaseline is a mixture of paraffin wax (a solid alkane) and oil.

The homologous series of alkenes

The alkenes are a series of unsaturated hydrocarbons, each containing a carbon–carbon double bond and having the general formula C_nH_{2n}, e.g.

ethene, C_2H_4 or $\overset{H}{\underset{H}{{}^{\diagdown}}}C=C\overset{H}{\underset{H}{{}^{\diagup}}}$.

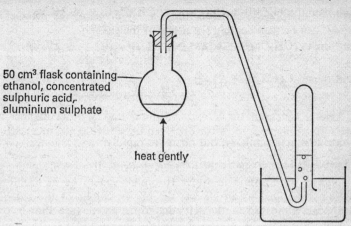

Apparatus for the preparation of ethene

Laboratory preparation of ethene (if needed)
The ethanol is dehydrated.

$$C_2H_5OH(l) \rightarrow C_2H_4(g) + H_2O(l)$$

Physical properties of ethene

Solubility in water	Colour	Odour	Density relative to air	Toxicity
Insoluble	Colourless gas	Virtually odourless	About the same	Non toxic

Chemical properties of ethene
As alkenes are unsaturated (unsaturated compounds contain at least one double or triple covalent bond), their reactions are often addition reactions.

*An **addition reaction** is when two or more molecules combine to form just one, larger molecule. The term is usually confined to organic chemistry.*

Addition with halogen elements If ethene is mixed with bromine vapour, a rapid addition reaction takes place. The colour of the bromine disappears, and sometimes a few drops of the product (a liquid) can be seen coating the inside of the gas jars.

$$C_2H_4(g) + Br_2(g) \rightarrow C_2H_4Br_2(l) \text{ (1,2-dibromoethane)}$$

or

$$\begin{array}{c} H \\ H \end{array} C=C \begin{array}{c} H \\ H \end{array} + Br_2(g) \rightarrow Br-\underset{\underset{H}{|}}{\overset{\overset{H}{|}}{C}}-\underset{\underset{H}{|}}{\overset{\overset{H}{|}}{C}}-Br$$

Chlorine reacts similarly, but far more rapidly.

Addition with solutions of the halogen elements If a small volume of a *solution* of bromine in water (bromine water) or in trichloromethane is added to an alkene, the brown colour of the halogen disappears almost instantly as it adds to the alkene (see above).

Note: Students are often confused by the fact that both alkanes and alkenes react with halogens, but by different reactions. The slow, multi-step *substitution* reactions of the alkanes must be contrasted with the rapid *addition* reactions of the alkenes.

Other addition reactions of ethene Ethene will react with hydrogen by addition to form ethane,

$$C_2H_4(g) + H_2(g) \xrightarrow[\text{high pressure, temperature}]{\text{nickel or platinum catalyst}} C_2H_6(g)$$

with hydrogen halides,

$$C_2H_4(g) + HCl(g) \rightarrow C_2H_5Cl(l) \text{ (monochloroethane)}$$

and with 'water' indirectly to form an alcohol, ethanol,

$$C_2H_4(g) + H_2O(g) \xrightarrow[H_3PO_4 \text{ catalyst}]{300°C, 60 \text{ atmospheres}} C_2H_5OH(g)$$

The polymerization of ethene and other alkenes is a special type of addition reaction, discussed in 12.4.

Reaction of ethene with acidified potassium manganate(VII) This reaction is not given by alkanes. If this reagent is added to an alkene, and the two substances shaken together, the purple colour of the manganate(VII) ion rapidly changes to a colourless solution. The reagent is a very powerful oxidizing agent, and oxidizes the alkene by an addition reaction, itself being reduced to the colourless Mn^{2+} ion.

Combustion of ethene Ethene and other alkenes burn in air, but as their molecules contain a higher proportion of carbon than do the

Organic chemistry 177

alkanes, they do not burn completely with a clean blue flame. The flame is yellow, and some smoke and soot are formed although most of the carbon does become carbon dioxide, e.g.

$$C_2H_4(g) + 3O_2(g) \rightarrow 2CO_2(g) + 2H_2O(g)$$

Uses of alkenes

Many alkenes are made by cracking petroleum, and their readiness to take part in addition reactions makes them very important starting materials for the manufacture of ethanol (see above), other alcohols, anti-freezes, coolants, detergents and (above all) polymers.

How to distinguish between methane and ethene

Test	Methane	Ethene
Combustion	Clean, blue, flame	Yellow flame, some soot
Bromine water	No reaction	Rapid decolorization
Acidified KMnO$_4$ solution	No reaction	Rapid decolorization

12.2 The oil industry

The separation of crude oil into fractions, both in industry and in the laboratory, has been discussed on pp. 40–43. Fractional distillation is only one of many processes carried out at an oil refinery, and *cracking* is particularly important.

There is an enormous demand for the gaseous and volatile liquid hydrocarbons from crude oil. These low molecular weight and low boiling point fractions are much used as liquid fuels (petrol, paraffin) and as raw materials for the manufacture of plastics, and so it is important to obtain them in high yields. However, large quantities of the heavier, less volatile fractions are also produced by the fractional distillation of crude oil, and although some of these have important uses (lubricating oils, etc.) they are not used on anything like the same scale as the simpler hydrocarbons.

The excess heavy fractions are not wasted, but rather reprocessed by cracking. This is essentially a brief exposure to a high temperature (e.g. by mixing with steam and heating to 900 °C for one second) which splits open the larger molecules into smaller molecules by breaking C—C bonds. This produces further supplies of the more simple hydrocarbons, e.g. for petrol. In addition, whenever a hydrocarbon molecule is cracked into two or more simpler molecules, at least one of the products is unsaturated. These small, unsaturated molecules such as ethene and propene are very reactive, and readily take part in polymerizations and other addition reactions to produce important chemicals. Cracking thus ensures much higher yields of important and reactive chemicals by reprocessing excess quantities of less important fractions (see Figure).

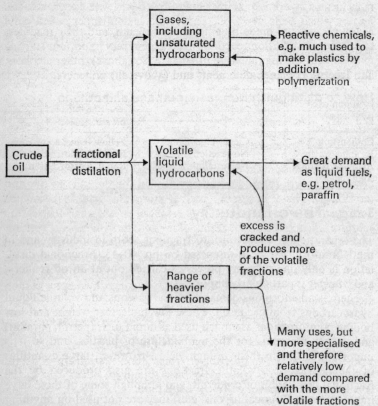

Refining crude oil

12.3 Alcohols, acids and esters

Alcohols
*A **functional group** is an atom or group of atoms present in an organic molecule and which tends to dominate the chemical properties of the molecule; the relatively inert carbon and hydrogen skeleton has no major influence on chemical properties.*

The alcohols are a series of compounds containing the —OH functional group and having the general formula $C_nH_{2n+1}OH$. The alcohols are named by replacing the –e of the appropriate alkane by the ending –ol, e.g. methanol, CH_3OH (from methane, CH_4), ethanol, C_2H_5OH and propanol, C_3H_7OH. The only alcohol studied in elementary work is ethanol, a colourless liquid with a characteristic odour, which boils at 78 °C. This liquid is sometimes called ethyl alcohol, or even just 'alcohol', which is particularly unfortunate as it is only one of many alcohols.

The laboratory preparation of ethanol by fermentation

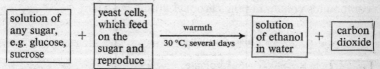

The yeast cells contain enzymes (biological catalysts) which break down the sugars to produce energy needed for the cells to live and reproduce. These enzymes, unlike inorganic catalysts, can be destroyed by over-heating so it is essential to keep the fermenting solution warm but not hot. Ethanol is produced when yeast respires anaerobically, i.e. breaks down sugars to produce energy without using oxygen. This is an alternative version of normal respiration, which is aerobic. Carbon dioxide and energy are formed in either case, but the sugar is not completely broken down into carbon dioxide in anaerobic respiration.

$$C_6H_{12}O_6(aq) \rightarrow 2C_2H_5OH(aq) + 2CO_2(g) \text{ (anaerobic)}$$
c.f. $\quad C_6H_{12}O_6 + 6O_2 \rightarrow 6CO_2 + 6H_2O \text{ (aerobic)}$

You could be asked to describe a fermentation experiment in detail, and how to obtain a reasonably pure sample of ethanol from the ferment by filtration followed by distillation.

Manufacture and uses of ethanol

Much of the world's supply of ethanol comes from fermentations of the above type, although the source of sugar or other form of carbohydrate may vary, e.g. potatoes, rice and sugar cane. The liquor from such distillations is used to produce various types of alcoholic drinks as well as ethanol itself. Another important manufacturing process is the hydration of ethene (p. 176) obtained from the cracking of petroleum. Large quantities of ethanol are used as a solvent and in the manufacture of other important organic chemicals.

Some chemical properties of ethanol
1 Burns with clean blue flame

$$C_2H_5OH(l) + 3O_2(g) \rightarrow 2CO_2(g) + 3H_2O(g)$$

(Methylated spirits largely consists of ethanol.)

2 Can be dehydrated by a variety of substances to form ethene, as on p. 175.
3 Reacts with organic acids, as in next section.

Organic acids (the carboxylic acids)
(Not all syllabuses include these compounds.) These are a series of compounds containing the carboxyl group, —COOH

$$\left(\text{also shown as } -C\overset{\displaystyle O}{\underset{\displaystyle OH}{\diagdown}}\right),$$

and having the general formula $C_nH_{2n+1}COOH$. The only one likely to be encountered at elementary level is ethanoic acid, CH_3COOH, commonly called acetic acid.

Physical properties of ethanoic acid The pure acid (glacial ethanoic acid) is a viscous liquid which freezes at 16 °C and then looks like ice. The acid has a characteristic smell of vinegar, and this persists in the dilute acid. Vinegar is largely dilute ethanoic acid.

Ethanoic acid as a typical acid The substance is an acid because it can donate a proton to a water molecule. It is important to realize that only *one* of the four hydrogen atoms in the molecule is 'acidic', and this is the one in the carboxyl group: $CH_3COO{\tiny\textcircled{H}}$.

$$CH_3COOH(l) + H_2O(l) \rightleftharpoons CH_3COO^-(aq) + H^+(aq)$$

The salts of the acid are called **ethanoates** (acetates). The acid shows all of the usual properties of a weak acid.

The reaction of organic acids with alcohols When an organic acid is heated with an alcohol in the presence of a little concentrated sulphuric acid as a catalyst, a reaction takes place to form an **ester**. Esters form a series of compounds containing the functional group

$$-\overset{\displaystyle O}{\underset{\displaystyle \|}{C}}-O-C-,$$

e.g. $CH_3COOH(l) + C_2H_5OH(l) \xrightarrow[\text{heat}]{\text{concentrated } H_2SO_4}$

$$CH_3COOC_2H_5(l) + H_2O(l)$$

Ethyl ethanoate (ethyl acetate) is the organic product, and it can also be shown as

$$CH_3-\overset{\overset{O}{\|}}{C}-O-C_2H_5.$$

The simple esters are colourless liquids with strong, pleasant odours, are immiscible with water, and tend to float on top of the reaction mixtures. They are used in perfumes and flavourings, and as solvents.

In the above reaction, the forward reaction is called **esterification** and the reverse reaction is a hydrolysis. Animal fats and vegetable oils are more complicated esters, formed by the combination of the trihydroxy alcohol glycerol,

$$\begin{array}{l} CH_2OH \\ | \\ CHOH\,, \\ | \\ CH_2OH \end{array}$$

and long chain carboxylic acids such as stearic acid. The hydrolysis of such fats by alkali produces the sodium salts of the acids (i.e. soap, p. 137) and free glycerol. This special type of hydrolysis is called **saponification**.

Note: Terylene is a polyester (p. 185).

12.4 Polymers

Polymers are macromolecules, i.e. very large molecules. Polymers are built up by the linking together of many smaller units called **monomers** to form much larger units which may consist of long chains, sheets or three-dimensional networks. This process is called **polymerization**. There are two main types of polymerization.

1 *Addition polymerization is the successive linking together of an unsaturated monomer. The only product is a single, large molecule. Each polymer normally contains only one kind of monomer molecule.*
2 *Condensation polymerization is the linking together of a large number of molecules by a reaction in which two products are formed, one of which is a small molecule (usually water). Such polymers may contain two different kinds of monomer molecule (e.g. in Terylene) or only one kind of monomer molecule (e.g. in starch).*

Some polymers occur in nature (natural polymers) and others are man-made (synthetic polymers). Artificial fibres are polymers which can be made into yarn for clothing.

*A **thermosetting polymer** softens when first heated but then undergoes a chemical change due to the formation of a network of cross linkages between polymer chains. This produces a rigid structure which cannot be softened or remoulded by subsequent heating.*

*A **thermoplastic** (**thermosoftening**) plastic softens on heating and hardens on cooling. This process can be repeated if necessary, and such materials are easily moulded into shape but are not heat resistant.*

Many polymers are extremely useful materials, for they each possess a range of properties unusual in one substance, e.g. strength, electrical insulation, waterproof, lightness, ease of moulding and resistance to corrosion.

Note: You will not need to learn all of the examples shown in Table 12.1 and in the description of condensation polymers: each syllabus has its own requirement.

Synthetic addition polymers
Many of these have a monomer with the same fundamental structure:

where X varies. For example, if X is a chlorine atom the monomer is vinyl chloride (chloroethene) and if X is a benzene ring (C_6H_5-) the monomer is phenylethene (styrene). See Table 12.1 for examples of addition polymers.

Some uses of these addition polymers

Polythene	Packaging, plastic bags, plastic film
Polyvinylchloride	Insulation for electric cables, raincoats, upholstery, suitcase coverings, gramophone records
Polystyrene	Heat insulator in buildings, packaging, model making. Expanded form made by generating gas in syrup during polymerization
Perspex	Used as a glass substitute, e.g. in aircraft windows, reflectors on vehicles, TV guard screens, etc.

Organic chemistry 183

Table 12.1 *Some synthetic addition polymers*

Name of polymer	Formula of monomer	Reaction conditions	Part of polymer chains
Polyethene (Polythene)	H₂C=CH₂ ethene (gas)	200 °C, 1000 atm, O_2 catalyst	$\left(-CH_2-CH_2-CH_2-CH_2-CH_2-\right)$ white, waxy solid
Polyvinylchloride (PVC) or polychloroethene	CH₂=CHCl vinyl chloride (chloroethene)	60 °C, high pressure, H_2O_2 catalyst	chain with alternating H and Cl on carbons
Polystyrene	CH₂=CH(C₆H₅) polyethene (styrene), syrupy liquid	Catalyst, e.g. dibenzoyl peroxide, heat	chain with C_6H_5 groups solid
Perspex	CH₂=C(CH₃)(COOCH₃) 'methyl-methacrylate', syrup	Catalyst, e.g. dibenzoyl peroxide, heat	chain with CH_3 and $COOCH_3$ groups transparent solid

Natural condensation polymers

Proteins from amino acids Amino acids have the general formula

$$NH_2-\underset{\underset{H}{|}}{\overset{\overset{R}{|}}{C}}-COOH,$$

where R is a side chain or hydrogen atom, which varies from one acid to another. NH_2 is the amino functional group. About 20 different amino acids occur in nature. Note that each molecule contains two different functional groups, NH_2 and COOH. (—[A]— and —[B]— are used below to signify the central, variable part of each monomer.)

NH_2—[A]—COOH + NH_2—[B]—COOH →
amino acid   amino acid

$$NH_2-[A]-CONH-[B]-COOH + H_2O$$
dimer

As the product still contains an NH_2 and a $COOH$ group, the process can continue at each end of the molecule, to build up a large molecule called a **protein**. Remember that a protein molecule will contain many different monomers, although each of the monomers will be an amino acid, and that each monomer could occur many times in the polymer.

Carbohydrates (starch) from sugars There are many different sugars, but only the polymers of glucose, $C_6H_{12}O_6$, are considered at an elementary level. For convenience, glucose is best considered as a 'block of atoms' (which takes no part in the polymerization) and 'reactive ends' of the molecule, which do.

OH—☐—OH + OH—☐—OH →
  two molecules of the monomer
  (a monosaccharide), glucose

OH—☐—O—☐—OH + H_2O
  dimer (a disaccharide),
  maltose

(The 'block' is $C_6H_{10}O_4$.) *Note* that as the dimer still contains 'reactive ends', the process can continue and a polymer (a polysaccharide) is built up, in this case starch.

An experiment to 'depolymerize' starch molecules You may have done an experiment which helped you to understand the structure of starch, in which case you could be asked to give experimental details. Briefly, a typical experiment involves the following principles.

1 If the polymerization process is a condensation (i.e. involves the elimination of water molecules) then the reverse is a hydrolysis, involving the addition of water molecules.
2 Starch solution is hydrolysed by acid hydrolysis (use dilute hydrochloric acid and boil) and/or by enzyme hydrolysis using enzymes in saliva at about 35 °C (starch is hydrolysed when it is digested).
3 Samples of the hydrolysate(s) are tested at intervals for starch (iodine solution goes blue-black) and for reducing sugars (boil with Fehling's A and B; brick red precipitate forms if reducing sugars present).
4 When hydrolysis is complete (i.e. no more starch present, only sugars) the hydrolysates are concentrated and chromatographed along with *known* samples of sugars such as glucose and maltose.

Organic chemistry 185

5 Development of the chromatogram shows that the acid hydrolysate contains glucose only (i.e. the monomer of starch is glucose, as explained above). The enzyme hydrolysate contains both glucose and maltose (a dimer of two glucose molecules) because this particular enzyme does not complete the digestion of starch.

Synthetic condensation polymers

Nylon This is similar to the polymerization of amino acids, but involves two different types of monomer, each of which contains two functional groups of the same type, e.g.

$$\text{COOH}-\square-\text{COOH} + \text{NH}_2-\square-\text{NH}_2 \rightarrow$$
$$\text{COOH}-\square-\text{CONH}-\square-\text{NH}_2 + \text{H}_2\text{O}$$

This process continues, to produce the polymer. The composition of the 'block' varies according to the type of Nylon, but typically contains 6 or 10 carbon atoms. You may have made a sample of Nylon in the laboratory. It is used mainly as a fibre, e.g. in clothing and ropes.

Terylene This is a polyester.

$$\text{COOH}-\square-\text{COOH} + \text{OH}-\square-\text{OH} \rightarrow$$
  'carboxylic acid'   'alcohol'

$$\text{COOH}-\square-\text{COO}-\square-\text{OH}$$
                    ester link

This process continues, to build up the polymer. It is used mainly as a fibre, e.g. in clothing and in fishing lines.

Questions

1 Alkanes: (a) are all gases; (b) have the general formula $C_nH_{2n+2}O$; (c) contain only carbon and hydrogen; (d) are usually soluble in water; (e) are usually reactive compounds.

2 A sample of oil was vaporized and the vapour passed over a heated catalyst; a gas formed which was found to decolorize bromine water. The gas: (a) is ethane; (b) contains ethane; (c) contains a saturated hydrocarbon; (d) contains an unsaturated hydrocarbon; (e) contains a mixture of alkanes.

3 The passage of gaseous ethene over a catalyst at high pressures to form a white flexible solid is called: (a) condensation; (b) neutralization; (c) esterification; (d) polymerization; (e) addition.

4 Large molecules can be built up by the combination of a number of smaller molecules. These smaller molecules are called: (a) polymers; (b) allotropes; (c) isomers; (d) dimers; (e) monomers.

5 A mixture of 400 cm³ of a gas A, and 400 cm³ of hydrogen reacted completely in the presence of a catalyst to give 400 cm³ of a new gas, B. When 400 cm³ of this new gas B were mixed with 400 cm³ of hydrogen in the presence of a different catalyst, 400 cm³ of ethane were produced. (All volumes measured under the same conditions of temperature and pressure.)

A liquid, D, was produced when B and bromine reacted together. Addition of another catalyst to B gave a solid, E, which did not react with bromine and had a molecular mass of about 20 000. Explain these results, identify A, B, D, E and give equations for the reactions.

How might polyvinyl chloride be obtained from A? (JMB)

6 Describe how you would obtain a sample of almost pure ethanol from sugar.

Outline the steps by which ethanol is manufactured from petroleum oil. Name all the chemical processes involved.

Describe *one* chemical and *one* physical test to distinguish between pure water and pure ethanol. State the result of each test on each of these substances. (JMB)

7 (a) Which homologous series of organic compounds can be represented by the following general formulae:

Series A $C_nH_{(2n+2)}$;
Series B C_nH_{2n};
Series C $C_nH_{(2n+1)}COOH$;
Series D $C_nH_{(2n+1)}OH$?

(b) Give the name and structural formula of *one* compound in each series.

(c) Describe reactions by which: (i) a *named* compound of series B can be converted to a compound of series A; (ii) a *named* compound of series D can be converted to a compound of series B.

(d) Name an important natural source of compounds of series A and give *two* industrial uses of such compounds. (CAM)

8 Chloroethane can be formed from an alkene by an addition reaction and from an alkane by a substitution reaction. Explain the meaning of the terms alkene, alkane, addition reaction, and substitution reaction and write equations for the two reactions referred to in the first sentence.

Describe (a) *two* other addition reactions of an alkene, and (b) *one* reaction by which chloroethane can be obtained from ethanol. (CAM)

13 Energy in chemistry

13.1 Energy changes

The following terms and ideas are essential for any understanding of energy changes in chemistry.

1. When chemicals react and give out energy (i.e. the container and its contents get hotter because of the reaction), the reaction is said to be **exothermic**. The opposite type of reaction is **endothermic**.
2. When chemical bonds are broken, energy is required to do this and the reaction is endothermic.
3. When chemical bonds are made, energy is released and the reaction is exothermic.
4. Energy changes are measured as the difference between the energy content of the starting materials (reactants) and that of the final materials (products) and are given the symbol ΔH.
5. In an exothermic reaction, when heat is given out, the reactants contain more energy than the products. As the energy content of the chemicals goes *down* during an exothermic reaction, ΔH in such cases is negative.

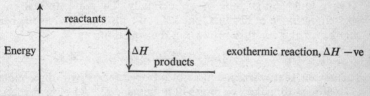

exothermic reaction, ΔH −ve

6. Conversely, endothermic reactions have a positive ΔH.

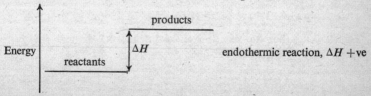

endothermic reaction, ΔH +ve

7 Energy changes are measured in joules or kilojoules (1 kJ = 1000 J). 4.18 joules are needed to raise the temperature of 1 g of water through 1 °C.
8 Most chemical reactions involve both bond-making (exothermic) and bond-breaking (endothermic) processes, and the overall ΔH for the reaction depends upon the magnitude of each of the steps.

There are many ways of describing ΔH, depending on the type of reaction concerned. The following include the more common examples.

Latent heat of vaporization

When water is boiled, its temperature does not change even though it is still being heated. This is because the molecules are being 'pulled apart from each other' in the liquid state and changed into free molecules in the gas state. This involves breaking intermolecular bonds, and the process is thus endothermic, i.e. requires heat energy. The energy needed to change liquid to vapour at the same temperature is called the **latent heat of vaporization**, and is often expressed in kJ per mole, i.e. kJ mol^{-1}. If the latent heats of vaporization of different liquids are compared (in kJ mol^{-1}), the figures are proportional to the strength of the intermolecular forces present. Water has a high latent heat of vaporization because it has relatively strong intermolecular forces, which require a lot of energy to break them.

Heat of reaction

This is the heat change involved with a particular chemical reaction, the reaction normally being stated in the form of an equation. Always look at the units of ΔH carefully. They are normally expressed per mole of product, but ΔH could also be given as the heat change for the quantities shown in the equation, e.g.

$$N_2(g) + 3H_2(g) \rightleftharpoons 2NH_3(g); \Delta H = -92.4 \text{ kJ}$$

This means that when one mole of nitrogen reacts with three moles of hydrogen to make two moles of ammonia, 92.4 kJ of heat energy are given out. The same information could be given as

$$N_2(g) + 3H_2(g) \rightleftharpoons 2NH_3(g); \Delta H = -46.2 \text{ kJ mol}^{-1}$$

This tells us that the reaction gives out 46.2 kJ of heat energy for every mole of ammonia formed.

Heat of precipitation

This is the heat change which occurs when one mole of a substance is precipitated from solution, e.g.

$AgNO_3(aq) + NaCl(aq) \rightarrow$
$\qquad AgCl(s) + NaNO_3(aq); \Delta H = -50.16$ kJ mol^{-1}

50.16 kJ is the heat of precipitation (of one mole) of silver chloride.

Such reactions are exothermic because strong bonds are being made (an ionic lattice) and all bond-breaking processes have occurred before reaction, when the reagents were dissolved in water. This is another example of where an ionic equation shows the only chemical reaction actually taking place,

$$Ag^+(aq) + Cl^-(aq) \rightarrow AgCl(s)$$

If you have done experiments to determine values for heats of precipitation, you could be asked to describe these, and to work out values from experimental results.

Suppose that 25.0 cm^3 of silver nitrate solution (concentration 0.5 mol dm^{-3}) are added to 25.0 cm^3 of sodium chloride solution (concentration 0.5 mol dm^{-3}) in an insulated container, and that there is a temperature rise of 3 °C. (If the container has a negligible heat capacity, e.g. a Polythene bottle, this simplifies the calculations.) 25 cm^3 + 25 cm^3 = 50 cm^3 solution.

Heat capacity of solution $\equiv$ heat capacity of water =
$\qquad\qquad\qquad\qquad\qquad\qquad\qquad\qquad$ 4.18 J g^{-1} °C^{-1}

50 cm^3 of solution rising through 3 °C has received $50 \times 4.18 \times 3$ J

This solution contains $\frac{25}{1000} \times 0.5$ mol of Ag$^+$ (or Cl$^-$), and therefore $\frac{25 \times 0.5}{1000}$ mol of silver chloride are precipitated = 0.0125 mol

$\therefore$ precipitation of 1 mole of silver chloride would produce
$\qquad \frac{1}{0.0125} \times 50 \times 4.18 \times 3$ J = 50 160 J or 50.16 kJ

The heat of precipitation of silver chloride = 50.16 kJ mol^{-1}

Note: If the container was made of a material which has a significant heat capacity (e.g. glass), then the heat absorbed by the container must be included in the calculation.

Heat of combustion

This is the heat change when one mole of a substance (in its normal state) is completely burned in oxygen. You may have done experi-

Heat of solution

These changes are not usually expressed in figures because the actual heat change depends not only on the amount of solid being dissolved, but also the volume of solvent it is dissolved in. It is important to understand, however, the processes which occur when a solid dissolves and the *type* of heat change which ensues.

Two basic steps are involved when an ionic solid dissolves in water.

1 The ions are separated from the crystal lattice, to become free ions in solution. This is a bond-breaking process, i.e. endothermic. Note that this process is described as dissociation of ions, *not* ionization, which has already occurred.
2 The separated ions react with water molecules to become hydrated ions, e.g. $Cu^{2+} + 4H_2O \rightarrow [Cu(H_2O)_4]^{2+}$, or, more simply, $Cu^{2+}(aq)$. This is a bond-making process, i.e. exothermic.

As (1) and (2) are opposites in their effect on ΔH, the overall ΔH for the reaction depends on the relative values of the two steps. If (2) is greater in magnitude than (1), then ΔH for the overall solution process will be exothermic. When anhydrous copper(II) sulphate dissolves in water, a great deal of energy is liberated and the reaction is exothermic. If hydrated crystals of the same salt are dissolved, however, the solution process is endothermic because in this case the ions are *already* hydrated, step (2) is relatively small in magnitude, and step (1) is the major contribution to the overall process.

Heat of neutralization

This is the heat change which occurs when a mole of oxide or hydroxide ions from a base is neutralized by hydrogen (hydronium) ions from an acid,

$$H^+(aq) + OH^-(aq) \rightarrow H_2O(l)$$

(Alternatively, this can be defined as the heat change when a mole of water is made by neutralization.)

You might expect that ΔH for these neutralizations will have the same value no matter which acid or which base is used, but this is not the case.

The neutralization of a strong acid by a strong alkali As strong acids and strong alkalis are completely dissociated in water, all of the

Energy in chemistry 191

bond-breaking processes have occurred before the chemicals react. The only change which then occurs is the bond-making process by which H^+(aq) and OH^-(aq) ions combine to form water molecules. The other ions present are merely spectator ions, and so *all* neutralizations between any strong alkali and any strong acid are exactly the same, and ΔH of neutralization has exactly the same value, i.e. -57.5 kJ mol^{-1} of H^+ ions (or OH^-, or H_2O).

Note: In calculations, be wary of reactions involving sulphuric acid, e.g.

$$H_2SO_4(aq) + 2NaOH(aq) \rightarrow Na_2SO_4(aq) + 2H_2O(l)$$

In this reaction, *two* moles of OH^- ions are neutralized by each mole of sulphuric acid. Make sure that ΔH is expressed *per mole* of OH^- (or H^+, or H_2O). The heat change for the reaction shown in the equation is -115.0 kJ, but the heat of neutralization per mole of H^+ (or OH^- or H_2O) is -57.5 kJ.

Neutralization of weak acids and alkalis Weak acids and alkalis are only partly dissociated in water and exist in equilibrium with their ions, e.g.

$$CH_3COOH(aq) \rightleftharpoons CH_3COO^-(aq) + H^+(aq)$$

In these solutions, all of the bond-breaking has *not* already occurred because some of the acid (or alkali) still exists as undissociated molecules. If a strong alkali (or acid) is added, H^+ (or OH^-) ions are neutralized and removed from the equilibrium to form water molecules. If this reaction were the only one taking place, ΔH of neutralization would again be -57.5 kJ mol^{-1} of hydrogen ions. However, as each ion is removed from the equilibrium mixture, more molecules dissociate to produce more of the ions, i.e. a bond-breaking step occurs. In addition, the 'new ions' appearing in the solution from this dissociation may at the same time become hydrated, which is another bond-making process. The neutralization reaction is thus modified by two other processes, one of which is bond-making and one of which is bond-breaking, and the overall ΔH depends upon these different contributions. ΔH values for neutralizations involving weak acids and/or alkalis thus vary in magnitude, according to the particular chemicals involved.

13.2 Fuels

A fuel is a substance used to provide energy. Most fuels are burned, and the energy produced in the combustion process is used as heat (e.g. from coal, natural gas and paraffin) or to provide the energy needed to operate machinery (e.g. to generate electricity, or in the internal combustion engine by 'exploding' petrol and air). The carbohydrates and fats that we eat are fuels which provide us with energy, and electricity is also a fuel.

The chemical energy within these fuels has normally been derived from the energy of the sun (e.g. coal and oil are indirectly produced by photosynthesis) and we are re-using this energy when we burn the fuels. Some of the relationships between fuels are summarized in the Figure.

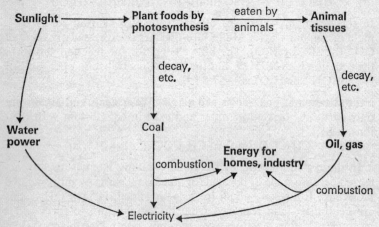

Some relationships between fuels

Comparing fuels The advantages and disadvantages of a particular fuel depend on answers to questions such as the following.

1. Is there a plentiful supply of the substance?
2. Can it be easily obtained, transported and stored?
3. Is it costly to purify so as to avoid impurities polluting the atmosphere when the fuel is burned (p. 81) and to avoid solid waste?
4. Does it have a 'high calorific value', i.e. does a given mass of it produce more energy than a similar mass of other fuels when it is completely burned?

   Note: When chemists compare substances, they normally take an equal number of particles of each, i.e. a mole, rather than equal masses. However, heats of combustion are sometimes quoted per

Energy in chemistry 193

unit mass or per unit volume of chemical, because fuels have to be transported. Transport costs are usually compared by considering the energy obtained from a given mass of fuel.

5 Is it (or its products) toxic?

A closer look at coal as a fuel Coal is produced from the remains of vegetable matter which lived millions of years ago and which has been subjected to pressure and bacterial action. Oil and natural gas are thought to be similar in origin, being derived from the remains of living creatures rather than vegetation.

Peat is a very early stage in the formation of coal, and different types of coal are named according to their properties and their 'age', e.g. lignite is a soft, crumbly, relatively 'young' form of coal, and anthracite is a hard, 'old' form.

Britain has larger stocks of coal than of any other fossil fuel, and coal is once again likely to increase in importance as a fuel. Large quantities of coal are burned to drive the turbines which generate electricity, and some coal is processed to produce coke and other solid smokeless fuels for industrial and domestic use. Traditionally, the destructive distillation of coal (heating coal in the absence of air, i.e. without burning it) was an important source of organic chemicals, fertilizers and coal gas, but crude oil has taken over as the main source of organic chemicals. The Haber process provides ammonia for fertilizers, and natural gas has replaced coal gas. Nevertheless, the destructive distillation of coal could achieve importance again as oil becomes expensive. You may have done a laboratory experiment to illustrate the process, in which case you could be asked to describe it. This processing of coal is summarized below.

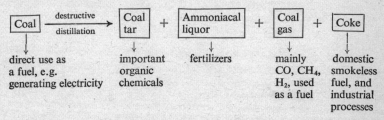

By altering the conditions, the balance of products may be changed, e.g. other solid smokeless fuels are obtained such as Coalite, Sunbright, etc.

Other fuels Most chemistry syllabuses are involved only with the study of oil, natural gas and (occasionally) coal as fuels, but physical science courses may also include the dry cell (battery), fuel cells and nuclear power, in which case you will need to revise the essential principles of these energy sources.

Questions

1 What is meant by the terms exothermic and endothermic reactions? Illustrate your answer by means of energy diagrams.

2 Define the term heat of neutralization. When a dilute solution of a strong acid is neutralized by a solution of a strong base, the heat of neutralization is found to be nearly the same in all cases. How can you account for this? How can you account for the cases where there is a difference?

3 Name two examples of: (a) a gaseous fuel; (b) a liquid fuel; and (c) a solid fuel. Can you think of any advantages or disadvantages each of these types has compared to each other?

4 (a) Describe and write equations for: (i) one reaction of industrial importance in which heat is absorbed; (ii) one reaction of industrial importance in which heat is produced; (iii) two reactions in which light energy is absorbed.

(b) What is meant by the calorific value of a fuel? A fuel such as coal and a food fuel such as bread are used in very different ways. Explain why the calorific value is an important characteristic of the fuel and food.

(c) Give the names and approximate composition (by volume) of two gaseous fuels.
(CAM)

5 Explain what is meant by an exothermic reaction. Give examples of exothermic reactions occurring between: (a) a gas and a solid; (b) a liquid and a solid; (c) two gases. State the conditions under which these reactions take place and the equation for the reaction.

Outline one commercial process in which the basic reaction of the process is exothermic.
(AEB)

6 (a) How do the processes of respiration and the burning of fuel (i) resemble one another; (ii) differ from one another?

(b) When we burn coal, we are making use of stored energy that came originally from the sun. Explain this statement.
(CAM)

14 Rates of reaction. Equilibria

14.1 Rates of reaction

Some chemical reactions proceed slowly (e.g. the rusting of iron) and others are extremely fast (e.g. explosions). Many of those reactions which are important in industry have reaction rates which can be increased with advantage, and so it is important to understand how such changes can be made.

In investigating the rate of a reaction, a rough guide can be obtained by noting the time taken for the reaction to go from start to finish. If the reaction is then repeated, using the same quantities but with a different set of conditions, and it is found to take less time from start to finish, it can be assumed that the rate of reaction has been increased. This approach is not very scientific, however, because it only gives an indication of the *average* speed; any reaction is proceeding faster at some stages than it is at others, and a more scientific approach is one which shows the actual rate of reaction at a given moment in time.

Rates are measured scientifically by finding how the concentration of a product increases with time, or (less likely) how the concentration of a reactant decreases with time. The units of rate of reaction, when measured in this way, are mol dm^{-3} s^{-1}.

How concentration changes affect reaction rates
The following experiments are typical of those used to investigate how concentration changes affect reaction rates.

1 The 'iodine clock', in which solutions of different concentrations are mixed to produce iodine, which then forms a blue colour in the presence of starch.
2 The precipitation of sulphur from sodium thiosulphate solution, by the addition of dilute hydrochloric acid; the concentration of the thiosulphate solution is varied. Typically, the reactions are

timed until enough precipitate has been formed to obscure a cross marked on a piece of paper below the reaction vessel.
3 The loss in mass when dilute hydrochloric acid solutions of differing concentrations are added to a fixed mass of calcium carbonate.

$$CaCO_3(s) + 2HCl(aq) \rightarrow CaCl_2(aq) + H_2O(l) + CO_2(g)$$

The loss in mass occurs because of the loss of carbon dioxide.

You could be asked to describe in detail one of these experiments. Remember that methods (1) and (2) above measure *overall* reaction times and only give an indication of average rates, whereas in (3) a graph is normally plotted (mass loss against time, as in the Figure), which indicates the actual rate at a given moment in time.

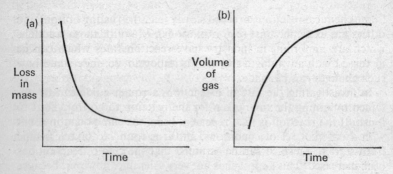

Variations of method (3) in which (a) the gas is allowed to escape and (b) the gas is collected in a syringe and its volume measured.

The effect of concentration changes on reaction rates

Remember also that in these experiments, and in all others in this section, every variable *except* the factor under investigation must be kept constant throughout the experiment. For example, in (3) above, each experiment must use the same *volume* of liquid, the same *mass* of calcium carbonate, and must be conducted at the same *temperature*; only the *concentration* of the acid solution can change in this particular investigation.

Experiments of this type show the following important points.

1 *A reaction rate increases if the concentration of one or more of the reactants is increased.*

   (*Note* that the concentration of a solid cannot be increased as its particles are already close together, but the concentration of a

Rates of reaction. Equilibria 197

solution can be increased by adding more solute, and that of a gas can be increased by increasing the pressure.)
2 The results are explained by the fact that increased concentrations cause a greater frequency of collisions between particles, and this increases the rate of reaction.
3 Another important point obtained from method (3) above is that the rate of a typical chemical reaction is fastest at the beginning (because the reagents are then at their highest concentrations) and gradually decreases (as substances are used up and their concentrations decrease), until it stops when one or more of the reagents is/are used up (see Figure).
4 Some students find this final point difficult to understand. In reactions of type (3), for a given amount of reagent the reaction will always produce the same loss in mass (or volume of gas), i.e. the graphs of the type shown in the Figure will always 'tail off' at the same level in a particular experiment. The *time* taken to reach this point may vary, however, according to the *concentration* of the substance. Thus if samples of 0.2 g of magnesium ribbon are added separately to 20 cm^3 of a hydrochloric acid solution of concentration 1 mol dm^{-3} (1 M) and then 20 cm^3 of hydrochloric acid of concentration 2 mol dm^{-3} (2 M), and the acid is in excess each time, the same volume of hydrogen will be formed in each case, but it will be formed more quickly with the more concentrated acid.

How temperature changes affect reaction rates

You could be asked to describe an experiment which investigates these changes. A typical one would be the reaction between sodium thiosulphate solution and dilute hydrochloric acid as used in (2) on p. 195. This time the *volume* and *concentrations* of the reagents are fixed, but the *temperature* is varied each time the experiment is done. Such experiments show that *the rate of a chemical reaction is increased if the temperature is increased.*

How surface area affects reaction rates

You could be asked to describe an experiment which investigates this, and a typical example would be one using hydrochloric acid and calcium carbonate as in (3) on p. 196. This time the *mass* of the solid, the *concentration* of the acid and the *temperature* are kept constant, but the solid is used first in lump form, then in smaller lumps and finally as a powder. Two points emerge from such experiments.

1 *If the surface area of a solid is increased (i.e. it is made more finely divided), the rate of reaction increases.*

2 This is explained by assuming that there is more contact between the reacting molecules when the surface area is increased. For example, most of the calcium carbonate in a lump cannot touch the acid because it is 'tucked away' inside the lump; when the lump is crushed, more of the carbonate can touch the acid.

How catalysts affect the rate of a reaction
A catalyst is a substance which alters the speed of a chemical reaction; it is unchanged chemically and in mass at the end of the reaction and so does not always appear to take part in the reaction, although it actually does so and is re-formed.

A positive catalyst speeds up a reaction. A negative catalyst slows down a chemical reaction.

You should know several examples of the use of a catalyst (e.g. some of the industrial processes in Unit 15, and in the laboratory preparation of oxygen, p. 84), and you could be asked to describe an experiment which shows that a chemical acts as a catalyst. A typical example would be the thermal decomposition of potassium chlorate(V) to produce oxygen, done with and without a catalyst such as copper(II) oxide,

$$2KClO_3(s) \rightarrow 2KCl(s) + 3O_2(g)$$

If you are asked to describe such an experiment, remember that you should use the results to indicate that the catalyst (1) alters the speed of the reaction and (2) is unchanged chemically and in mass at the end of the reaction. Many students forget the (2) part of the argument.

Remember also that enzymes are a special type of catalyst found in living organisms. These show the same general properties of any catalyst, but they are destroyed by heating, unlike ordinary catalysts.

How light affects the rate of a reaction
This is the least important of the factors considered, because whereas the other factors apply in nearly all cases, light only noticeably affects a few reactions. Nevertheless, one of these, photosynthesis, is arguably one of the most important reactions in the world. This reaction also serves to remind us that light is a form of energy, and therefore its influence on some reactions should not be a surprise.

You may know that a mixture of hydrogen and chlorine will

Rates of reaction. Equilibria 199

explode if subjected to light, and you may have studied the effect of light on the halides of silver (AgCl, AgBr, AgI), which is utilized in photographic films and papers.

14.2 Reversible reactions and equilibria

Reversible reactions

A reversible reaction is one which can proceed both from left to right (as shown by an equation) and also from right to left. They are indicated by using the sign $\rightleftharpoons$ (rather than $\rightarrow$) in the equation. You will have encountered many reactions of this type, some of which are listed below. Remember that *everything* is reversible in such reactions, including energy changes; if heat is given out (ΔH —ve) when the reaction goes from left to right, then the reverse reaction will take in heat (ΔH +ve). By convention, the ΔH symbol given with any equation for a reversible reaction refers to the forward reaction only, i.e. the reaction from left to right as shown in the equation. The back reaction will have a ΔH of the same magnitude, but opposite in sign.

$$5H_2O(l) + CuSO_4(s) \underset{\text{endothermic}}{\overset{\text{exothermic}}{\rightleftharpoons}} CuSO_4 \cdot 5H_2O(s)$$

$$CaCO_3(s) + H_2O(l) + CO_2(g) \underset{\substack{\text{softening temporary hard} \\ \text{water, formation of} \\ \text{kettle fur, etc.}}}{\overset{\text{formation of hard water}}{\rightleftharpoons}} Ca(HCO_3)_2(aq)$$

Equilibrium

Imagine a reversible reaction $A + B \rightleftharpoons C + D$, starting with only A and B, and allowed to continue without interference. When A and B are mixed, the forward reaction is at first rapid, because the concentrations of A and B are at their highest. At the same time, the back reaction is very slow because there are virtually no molecules of C and D. As time goes on the rate of the forward reaction decreases (as A and B are used up) and the rate of the back reaction increases (as more of C and D are formed) until the point is reached when the rate of both reactions is the same. This situation is then described as an equilibrium (balance point), and there will be no further change in the concentrations of the various chemicals present unless the conditions are changed, e.g. unless more reagents are added, or something is removed, or the pressure is changed, etc.

Chemical equilibria are dynamic, i.e. both reactions still continue (at equal rates) at equilibrium although they do not appear to be continuing as there is no further change in the concentrations. This can be shown by the fact that a saturated solution of lead(II) chloride becomes radioactive when placed in contact with a sample of radioactive (solid) lead(II) chloride. The saturated solution cannot dissolve any more lead(II) chloride, but it 'swaps' dissolved lead ions in the solution for radioactive lead ions in the solid, thus showing that the equilibrium $Pb^{2+}(aq) \rightleftharpoons Pb^{2+}(s)$ is dynamic. An equilibrium would be static if both reactions stopped at equilibrium.

Students frequently imagine that equilibrium is attained when equal concentrations of reactants and products are obtained. This is very rarely the case, and equilibrium positions vary enormously, e.g. it is possible to have a situation consisting of 99% products and 1% reactants, but another equilibrium may contain 1% products and 99% reactants.

How equilibrium positions can be changed

If a chemical is being manufactured by a reversible reaction, an equilibrium mixture may contain only a small proportion of the desired product. It would obviously be very useful if the equilibrium position could be changed so as to produce a new equilibrium position richer in the desired chemical. Remember that the only way in which this can be achieved is by making a change in the *conditions* of the process.

In 14.1 five factors were considered, each of which can influence the rate of a reaction. These factors still apply to reversible reactions, and they apply to *both* reactions in a reversible reaction. However, a change in one of these factors can influence one of the two reactions in a reversible reaction more than the other, and this is how an equilibrium position can be changed. The changes (explained more fully in the examples which follow) can be predicted by using the **Principle of Le Chatelier**: 'If a change is made to a system in equilibrium, the system alters so as to oppose the change, and a new equilibrium is formed.' This can be rephrased in simple practical terms as 'whatever is done to a system in equilibrium, the system does the opposite.' *Learn the Principle*; you do not have to understand why the changes occur.

How concentration changes affect a system in equilibrium Consider the equilibrium $A + B \rightleftharpoons C + D$. If more A (or B) is added at equilibrium, the rate of the forward reaction will be increased (because there will be more collisions between A and B) but the back reaction will not be affected as it does not involve collisions between molecules of A and B. A new equilibrium position will be reached, richer in products. Similarly, if C (or D) is removed, the rate of the back reaction decreases and more products are made. Each of these can be predicted by the Principle of Le Chatelier, for whatever change is made (e.g. adding A) the system does the opposite (e.g. removes A).

If reactants are added or products are removed in an equilibrium, a new equilibrium is produced which contains a higher proportion of the products.

You may have done an experiment to illustrate this idea, e.g. the reaction between bismuth(III) chloride and water:

$$BiCl_3(aq) + H_2O(l) \rightleftharpoons BiOCl(s) + 2HCl(aq)$$

Alternative additions of water and concentrated hydrochloric acid force the equilibrium first one way and then the other, and the changes can be observed because of the appearance and disappearance of the precipitate.

Reversible changes occur naturally. The reaction

$$CaCO_3(s) + H_2O(l) + CO_2(g) \rightleftharpoons Ca(HCO_3)_2(aq)$$

does not reach equilibrium in a cave system where the solution is constantly removed in the streams which leave the cave; the reaction in such circumstances is constantly driven from left to right and the calcium carbonate is continually 'dissolved'. On the other hand, if temporary hard water (e.g. $Ca(HCO_3)_2$ solution) is boiled in a kettle or allowed to evaporate, the same reaction is forced the other way because the CO_2 and H_2O are constantly escaping. This produces kettle fur in kettles or stalactites and stalagmites in caves.

How temperature changes affect a system in equilibrium An increase in temperature in a reversible reaction will increase the rates of both forward and backward reactions, and so equilibrium is attained more quickly. There is also an additional effect, which changes the equilibrium position, because one of the reactions is speeded up more than the other.

When the temperature of a system in equilibrium is raised, a new equilibrium is formed which contains a higher proportion of the material(s) made by the endothermic reaction (ΔH +ve). A similar change in favour of the exothermic process is produced by a lowering of the temperature.

The above statements can again be predicted by the Principle of Le Chatelier, for if the temperature is *increased* the system tries to *decrease* it, and this is achieved by increasing the rate of the endothermic process, which 'absorbs' some of the extra heat energy.

How catalysts affect a system in equilibrium If a positive catalyst is used in a reversible reaction, the rates of *both* forward and backward reactions are increased *equally*.

A catalyst does not alter the equilibrium position (i.e. the yield) but it ensures that equilibrium is reached more quickly than without it; this is an important economic consideration.

How pressure changes affect a system in equilibrium Pressure changes can only affect reactions between *gases*, for although gases can be compressed (i.e. concentrated), solids and liquids cannot. An increase in pressure has the same effect as increasing the concentration of a gas, so that both forward and backward reactions involving gases are speeded up and equilibrium is reached more quickly. There can also be a second effect, which is usually more important, when one of the reactions is speeded up more than the other and a new equilibrium is formed.

If the equation for the reaction shows a different number of gas molecules on the left hand side than there are gas molecules on the right hand side, then a change in pressure will produce a different equilibrium position.

If the pressure is increased, the system tries to decrease the pressure (Principle of Le Chatelier) by increasing the rate of the reaction which produces fewer molecules, e.g.

$$2A(g) + B(g) \rightleftharpoons 2C(g)$$
(3 molecules of gas)   (2 molecules of gas)

Reaction → produces fewer molecules, so increasing the pressure increases the rate of → more than that of ←, and a new equilibrium is formed with a higher proportion of C.

A summary of the factors which affect equilibrium
1 For a reaction

$$A_2(g) + B_2(g) \rightleftharpoons 2AB(g); \Delta H = -210 \text{ kJ mol}^{-1},$$

the following changes would produce a new equilibrium with a higher yield of product:
(*a*) lowering the temperature, (*b*) adding more of A and/or B, and

Rates of reaction. Equilibria 203

also removing AB. Pressure changes will not affect the yield. The use of a positive catalyst will not increase the yield, but will reduce the time taken to reach equilibrium.

2 In the reaction

$$D_2(g) + 3E_2(g) \rightleftharpoons 2DE_3(g); \Delta H = +810 \text{ kJ mol}^{-1},$$

the following changes would produce a new equilibrium with a higher yield of product:
(a) raising the temperature (particularly useful because it also reduces the time taken to reach equilibrium), (b) adding more of D_2 and/or E_2, and also removing DE_3, (c) raising the pressure. A positive catalyst will not increase the yield, but it is important because it reduces the time taken to reach equilibrium.

The conditions used for industrial processes are often decided by a consideration of the above factors. However, the theoretical conditions may have to be modified for economic reasons, e.g. a lowering of the temperature may in theory produce a higher yield, but this will also result in the slowing down of both reactions and a longer time to reach equilibrium, and a compromise temperature may have to be used.

Questions

1 (a) The formation of methanol (methyl alcohol) from hydrogen and carbon monoxide can be represented by
$$CO(g) + 2H_2(g) \rightleftharpoons CH_3OH(l), \Delta H = +91 \text{ kJ mol}^{-1}$$
What mass of hydrogen would react to cause a heat change of 91 kJ?
(b) What would be the effect on the equilibrium concentration of methanol in this endothermic reaction if (i) the temperature was increased; (ii) the pressure was increased; (iii) the hydrogen concentration was increased? (JMB)

2 Explain why an increase in the concentration of one or more of the reactants increases the rate of a chemical reaction.

3 An increase in pressure does not noticeably affect the rate of a reaction between solids but increases the rate of a reaction between gases. Explain.

4 Distinguish between the terms static and dynamic as applied to equilibria.

5 The equation for the reaction by which ammonia is manufactured is:
$$N_2(g) + 3H_2(g) \rightleftharpoons 2NH_3(g)$$
(a) What would be the effect on the equilibrium concentration of ammonia of (i) increasing the pressure; (ii) increasing the nitrogen concentration?
(b) The equilibrium concentration of ammonia increases as the temperature is lowered. Is heat evolved or absorbed when ammonia is formed?
(c) Why is a catalyst used in this reaction? (JMB)

6 $2SO_2(g) + O_2(g) \rightleftharpoons 2SO_3(s)$ + heat evolved, $\Delta H = -189 \text{ kJ mol}^{-1}$
The equation represents a system in equilibrium. State the changes in the equilibrium concentration of sulphur trioxide which would be caused by: (a) adding oxygen; (b) heating the mixture; (c) increasing the pressure. (JMB)

7 In the reaction
$$2Y + W \rightleftharpoons Y_2W, \Delta H = -8400 \text{ kJ mol}^{-1}$$
which of the following would result in a higher yield of Y_2W?
   (a) The use of a suitable positive catalyst.
   (b) Lowering the temperature.
   (c) Removal of W.
   (d) Reducing the pressure.
   (e) Increasing the surface area of W.

8 Explain what is meant by the term catalyst. Describe a simple experiment or experiments to demonstrate the use of a catalyst. Give two industrial reactions in which a catalyst is used, naming the initial substances, the product(s) and the catalyst used as well as any essential conditions. (Long accounts are not wanted, nor are diagrams. The manufacture of nitric acid is excluded.) (CAM)

9 The following statements are made in a textbook. 'The rates of most chemical reactions are approximately doubled by raising the temperature at which the reactions are carried out by 10 °C.'

'The rate at which a chemical substance reacts is directly proportional to its concentration.'

(a) Describe the experiments you would carry out to test the truth of these two statements when applied to either the reaction between a metal and a dilute acid or the decomposition of hydrogen peroxide catalysed by manganese dioxide (manganese(IV) oxide).

(b) Explain simply, in terms of the ions or molecules present, why the rate of a reaction is increased both by raising the temperature and also by increasing the concentration of the reagents. (CAM)

15 The chemical industry

15.1 Social and economic factors

The siting of a particular factory and the conditions used to operate the chemical reactions within it are decided by social and economic factors. It is possible that an examination question could ask you to justify the actual *conditions* used to operate a chemical process (summarized on p. 202). Sometimes it is possible to make a chemical by more than one method, in which case the choice may depend upon whether any side products are of commercial value. The choice of a particular *site* for a factory may depend upon points such as the following.

1. Are the raw materials readily available in the locality, and if not can they be transported easily to the factory?
2. Is the finished product utilized in the locality, or if not is the factory well served by transport systems (e.g. motorways, ports, etc.) to facilitate distribution?
3. If large quantities of a particular fuel are needed (e.g. coal, oil) are these readily available in the locality?
4. If large quantities of water are needed for cooling, etc. is the local water soft?

If such factors are included in your syllabus, you will probably have discussed other examples of this type.

15.2 Common industrial processes

Note: You may not need to study all of these reactions for the particular syllabus you are following.

The production of ammonia by the Haber process

$$N_2(g) + 3H_2(g) \rightleftharpoons 2NH_3(g); \Delta H = -46.2 \text{ kJ mol}^{-1}$$

Gases: purified and mixed in the ratio 1:3 (nitrogen from the air, hydrogen from natural gas).
Pressure: as high as economically possible (Le Chatelier's Principle, hereafter referred to as LCP), typically 300 atmospheres.
Temperature: theoretically low (LCP), but reaction rate would be too slow even with a catalyst, so typical optimum temperature is 400 °C.
Catalyst: finely divided iron, mixed with other chemicals to improve its action. This does not increase the yield, but decreases the time taken to reach equilibrium.
Removal of ammonia: the product is removed by liquefaction or by dissolving in water.

The manufacture of sulphuric acid by the Contact process

$$2SO_2(g) + O_2(g) \rightleftharpoons 2SO_3(g); \Delta H = -94.9 \text{ kJ mol}^{-1}$$

Gases: sulphur dioxide from the combustion of sulphur, oxygen from liquid air. (These *must* be pure and dry.)
Pressure: theoretically as high as economically convenient (LCP), but in practice atmospheric pressure is used as even then the equilibrium position is already well over to the right.
Temperature: theoretically low (LCP), but reaction rate would be too slow even with a catalyst, so typical optimum temperature is 450° C.
Catalyst: vanadium(V) oxide, which does not produce a higher yield but reduces the time taken to reach equilibrium.
Removal of sulphur trioxide: product dissolved in fairly concentrated sulphuric acid (p. 153) which is thus made even more concentrated. Some of this is removed as product, and the rest is diluted and used to absorb more sulphur trioxide. In effect:

$$SO_3(g) + H_2O(l) \rightarrow H_2SO_4(aq)$$

The extraction of aluminium metal, sodium metal and sodium hydroxide by electrolysis
See p. 52, p. 54 and p. 56.

The production of iron in the blast furnace
1 *Raw materials:* iron ore (impure haematite, iron(III) oxide, Fe_2O_3), coke and limestone (all added at the top of the furnace, see Figure) and oxygen from air 'blasted' in at the bottom of the furnace.

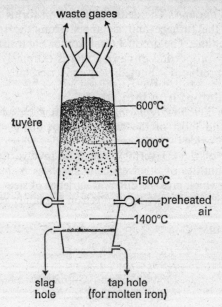

The blast furnace

2 *Chemical reactions producing iron:* hot coke burns to form carbon dioxide (more carbon dioxide is also obtained from heating limestone as in (3) below),

$$C(s) + O_2(g) \rightarrow CO_2(g)$$

The carbon dioxide then reacts with more hot coke to form carbon monoxide,

$$CO_2(g) + C(s) \rightarrow 2CO(g)$$

The carbon monoxide reduces the iron ore to iron,

$$Fe_2O_3(s) + 3CO(g) \rightarrow 2Fe(l) + 3CO_2(g)$$

3 *Removal of waste products:* Waste gases (mainly nitrogen and oxides of carbon) escape at the top. The limestone decomposes to form calcium oxide,

$$CaCO_3(s) \rightarrow CaO(s) + CO_2(g)$$

which reacts with the silicon impurities present in the ore to produce slag (molten calcium silicate) which floats on top of the molten iron.

$$CaO(s) + SiO_2(s) \rightarrow CaSiO_3(l)$$

The slag is 'tapped off' from time to time.

4 *Removal of the iron:* The dense molten iron forms a layer at the bottom of the furnace and is tapped from the furnace at the appropriate time. The product (cast iron or pig iron) still contains a significant proportion of carbon and is extremely brittle. Most of the iron produced is immediately converted into steel.

The conversion of iron into steel

Principle This involves *removing* impurities such as sulphur and phosphorus and most of the carbon (by oxidizing them to form either volatile oxides which escape, or oxides which are absorbed by the lining of the convertor), and may involve the *addition* of controlled quantities of other elements, such as manganese, cobalt and tungsten, which give the different varieties of steel their different properties. The excess carbon in pig iron is mainly responsible for

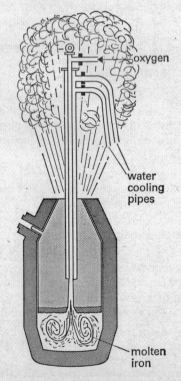

The LD process *(British Iron and Steel Federation)*

its brittle nature, but a small controlled proportion of carbon is an essential constituent of steel.

A typical process: the LD process The LD convertor (see Figure) is tilted to receive the charge of molten iron, and a water-cooled steel lance blows oxygen at 5 to 15 atmospheres pressure on to the surface. This 'burns out' the impurities as volatile oxides or as oxides which (being acidic) are absorbed by the basic lining of the convertor.

The manufacture of sodium carbonate by the Ammonia-Soda (Solvay) Process

Raw materials: sodium chloride (as brine), limestone, coal and ammonia.

Chemical reactions: (the diagram shows a flow diagram for the process.)

The limestone is heated in a kiln:

1 $CaCO_3(s) \rightarrow CO_2(g) + CaO(s)$

The carbon dioxide is passed up the Solvay tower (see diagram)

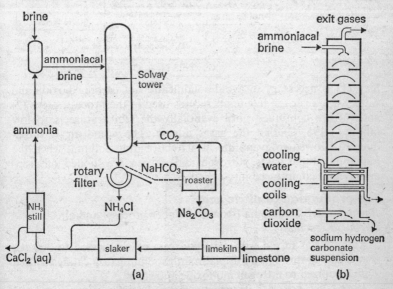

(a) A flow diagram of the ammonia—soda or Solvay process
(b) The Solvay tower

where it meets the ammonia dissolved in the brine (ammoniacal brine) trickling down the tower:

2 Na$^+$(aq) + Cl$^-$(aq) + CO$_2$(g) + NH$_3$(aq) + H$_2$O(l) →
$$\text{NaHCO}_3(s) + \text{NH}_4^+(aq) + \text{Cl}^-(aq)$$

The precipitated sodium hydrogencarbonate is filtered off on a rotating filter, washed and roasted in a furnace:

3 2NaHCO$_3$(s) → Na$_2$CO$_3$(s) + CO$_2$(g) + H$_2$O(l)

Economics of the process: This is an outstanding example of economy in the use of materials. The carbon dioxide from reaction (2) is used to supplement that provided by reaction (1). The calcium oxide from (1) is 'slaked' to form the hydroxide,

$$\text{CaO(s)} + \text{H}_2\text{O(l)} \rightarrow \text{Ca(OH)}_2(s)$$

which, being a base, can be heated with the ammonium chloride from reaction (2) to form more ammonia. The only waste product is calcium chloride. The main product, sodium carbonate, is an essential raw material for a large number of industries.

The liquefaction of air

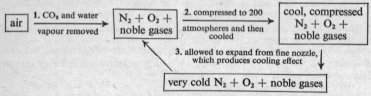

Step (1) is necessary to avoid solidification of carbon dioxide and water vapour at the low temperatures used in the process. Steps (2) and (3) are continued until eventually the temperature is so low (about −195 °C) that the gases liquefy. The liquid air is subsequently allowed to evaporate in a fractionating column, when the individual noble gases, nitrogen and oxygen are 'boiled off' separately and compressed into cylinders.

The production of nitric acid

Raw materials: ammonia (from the Haber process) and air.
Chemical reactions:

1 Ammonia is mixed with a large excess of air, heated to 300 °C, and passed through a hot platinum/rhodium gauze catalyst, when it is oxidized to nitrogen monoxide and steam.

$$4\text{NH}_3(g) + 5\text{O}_2(g) \rightarrow 4\text{NO}(g) + 6\text{H}_2\text{O}(g); \Delta H = -1807 \text{ kJ mol}^{-1}$$

The exothermic reaction, once started, provides the heat necessary for the oxidation to continue.

2 The hot gases are used to preheat more ammonia, and are cooled in the process. At the lower temperature, the nitrogen monoxide combines with oxygen in the air to form nitrogen dioxide,

$$2NO(g) + O_2(g) \rightarrow 2NO_2(g)$$

3 The gases pass into a series of absorption towers where they meet a stream of dilute nitric acid or water, and dissolve to form nitrous and nitric acids,

$$2NO_2(g) + H_2O(l) \rightarrow HNO_3(aq) + HNO_2(aq)$$

4 The nitrous acid is oxidized by excess oxygen to nitric acid, so that the nitric acid concentration in the liquids in the towers gradually increases. Acid of about 55–65% concentration is eventually obtained.

Questions

1 The object of a diaphragm in the diaphragm cell for the electrolysis of sodium chloride solution is: (*a*) to prevent production of hypochlorite; (*b*) to assist production of hypochlorite; (*c*) to keep up the resistance of the cell; (*d*) to prevent depletion of the solution around the anode; (*e*) to prevent a rise in temperature.

2 The raw materials fed into the blast furnace for making iron are: (*a*) iron(II) oxide, calcium carbonate and coke; (*b*) iron(III) oxide, calcium oxide and coke; (*c*) iron(III) oxide, calcium carbonate and coke; (*d*) tri-iron tetroxide, calcium hydroxide and coke.

3 Answer the following questions about the manufacture of iron and steel (no diagrams are required).

(*a*) Give the name and formula of one mineral from which iron is extracted.

(*b*) Explain how carbon monoxide is formed in the blast furnace.

(*c*) Write the equation for one reaction by which metallic iron is formed in the furnace.

(*d*) Explain clearly why limestone (calcium carbonate) is used in the blast furnace and suggest what you think would happen if the limestone were not present.

(*e*) Name three impurities likely to be present in the 'pig iron' formed in the blast furnace. Give one effect of these impurities on the physical properties of the iron.

(*f*) Explain how these impurities are removed during the conversion of pig iron into steel. (CAM)

4 Describe carefully how sodium carbonate is manufactured by the ammonia–soda (Solvay) process. No details of the plant are required. Give two reasons why this is a very economical process. Explain briefly how glass is manufactured. (JMB)

5 Sulphuric acid is manufactured by converting sulphur dioxide to sulphur trioxide and dissolving this in 95–98% acid, while adding water to the appropriate extent.

(*a*) How is the sulphur dioxide obtained?

(*b*) State the approximate temperature and relative proportions of gases used in the conversion.

(*c*) Name one of the catalysts commonly used.

(*d*) Explain why one uses a moderate temperature and a catalyst.

(*e*) Why is the sulphur trioxide not dissolved in water directly?

(*f*) State two important commercial uses of sulphuric acid.

Describe one reaction in each case in which sulphuric acid acts as (i) an acid; (ii) a dehydrating agent; (iii) an oxidizing agent. (AEB)

16 More calculations

16.1 Faraday's laws of electrolysis

The following information is essential to the understanding of Faraday's laws.

Quantity of electricity is measured in coulombs. Number of coulombs used = current in amps × time of current flow in seconds.

*One **faraday** of electricity (96 500 C) consists of a mole of electrons.*

One faraday is needed to discharge a mole of a monovalent ion during electrolysis, two faradays for a mole of a divalent ion, and three faradays for a mole of a trivalent ion.

Faraday's first law of electrolysis
This law states that the mass of a substance dissolved off, or produced at, an electrode during electrolysis is proportional to the quantity of electricity which passes through the electrolyte.

Thus if 300 coulombs deposit 0.1 g of copper at a cathode, then 600 coulombs will deposit 0.2 g of copper from the same electrolyte. The results are sometimes expressed graphically, in which case a plot of quantity of electricity against mass of product will be a straight line.

Some syllabuses could require you to describe an experiment which illustrates this law, in which case remember to include an ammeter and a variable resistance (so that the current can be kept constant) in the circuit.

Faraday's second law of electrolysis
This states that the number of moles of different elements released by the same quantity of electricity are in simple ratios. (Remember that experimental results will be given in masses, and these need to be converted into moles in order to verify the law.)

More calculations 213

This law is best understood by considering the results obtained by passing the same current for the same length of time through a number of different electrolytes. The following is a typical set of results for masses of elements deposited/evolved at the cathode from different electrolytes, e.g. silver from silver nitrate solution.

silver	*magnesium*	*hydrogen*	*lead*	*aluminium*
0.535 g	0.060 g	0.005 g	0.515 g	0.045 g

Dividing the mass of each metal or hydrogen by its atomic mass, so converting it into a number of moles of atoms:

$\dfrac{0.535}{108}$	$\dfrac{0.060}{24}$	$\dfrac{0.005}{1}$	$\dfrac{0.515}{207}$	$\dfrac{0.045}{27}$
=0.005	0.0025	0.005	0.0025	0.0017

The number of moles of silver and hydrogen are the same, and twice that obtained for magnesium and lead, and three times the number of moles of aluminium. This is explained by the fact that a mole of aluminium ions needs three faradays to be discharged (the ion is Al^{3+}), whereas a mole of silver ions (Ag^+) needs only one faraday. It thus follows that the number of atoms of silver liberated by a certain quantity of electricity will be three times the number of atoms of aluminium deposited by the same quantity of electricity.

Calculations using Faraday's laws
Two typical examples are given.

(1) The same current is passed for the same time through a solution of silver nitrate and through a separate voltameter containing dilute sulphuric acid. If 0.108 g is deposited on the silver cathode, what mass of hydrogen gas will be liberated?

Ag^+ and H^+ are both univalent, so the same quantity of electricity will liberate an equal number of moles of atoms of each.

0.108 g of silver is $\dfrac{0.108}{108}$ moles of silver = 0.001 moles.

∴ 0.001 moles of hydrogen atoms (*not* molecules) will also be liberated.

0.001 moles of hydrogen atoms = $\dfrac{0.001}{1}$ g = 0.001 g.

(2) Calculate the mass of magnesium produced by the electrolysis of fused magnesium chloride, if a current of 1.93 A is passed for 16 minutes and 40 seconds.

16 minutes and 40 seconds = 1000 seconds.

Quantity of electricity used $= 1000 \times 1.93$ coulombs $= 1930$ C

1930 C $= \dfrac{1930}{96\,500}$ faradays

Two faradays will discharge one mole of magnesium ions, Mg^{2+}, i.e. 24 g

∴ $\dfrac{1930}{96\,500}$ faradays will discharge $\dfrac{1930}{96\,500} \times 12$ g magnesium.

∴ Mass of magnesium liberated $= 0.24$ g

16.2 Volume changes in gases

Standard temperature and pressure

Boyle's law (p. 18) and Charles' law (p. 17) lead to the general gas equation

$$\frac{P_1 V_1}{T_1} = \frac{P_2 V_2}{T_2}$$

This equation enables us to change the volume of a gas measured under a certain set of conditions to that it would occupy under a different set of conditions.

Standard temperature and pressure (stp) defines the conditions at which most gas volumes are compared, and these are a temperature of 273 K (0 °C) and a pressure of 760 mm of mercury (1 atmosphere).

Suppose that the volume of a gas is 300 cm³ at 25 °C and 780 mm mercury pressure, and we are asked to determine what its volume would be at stp.

$$\frac{P_1 V_1}{T_1} = \frac{P_2 V_2}{T_2}, \quad \therefore V_2 = \frac{P_1 V_1 T_2}{P_2 T_1}, \quad \therefore V_2 = \frac{780 \times 300 \times 273}{760 \times 298}$$

(remember that T is always in K, not °C).

∴ $V_2 = 282$ cm³.

Gay Lussac's Law

Gay Lussac's law states that when gases combine, they do so in volumes which bear a simple ratio to one another and to the volume of the products (if gaseous), all volumes being measured at the same temperature and pressure.

You may have seen experiments to illustrate this law, using for example 100 cm³ glass syringes. Alternatively you could be provided with experimental data and asked to use it to illustrate the law. The essential things to remember in performing these calculations are given below.

1 Use only the volumes of gases which are *used up* or *formed* during the reaction. Some gases may be in excess, and the excess should not enter into the calculation.
2 If steam is one of the products, and the measurements are made at room temperature or below, the steam will condense to a *negligible* volume of liquid water, and this should not enter into the calculations.
3 If more than one gas is formed, it is sometimes necessary to find the volumes of the individual gases by absorbing one of them in a solution (e.g. carbon dioxide can be absorbed in potassium or sodium hydroxide solution). The volumes before and after absorption indicate the volume of the gas which has been absorbed.

The following example should make this clear.

48 cm^3 of methane were mixed with 212 cm^3 (an excess) of oxygen, and the mixture was exploded. The products, after cooling and reverting to room temperature and pressure, occupied 164 cm^3 which became 116 cm^3 after shaking with sodium hydroxide solution. We are asked to show that these results are in accordance with Gay Lussac's law.
Methane + oxygen → carbon dioxide + steam (which condenses to a negligible volume of water).
The oxygen is in excess, so all the methane is used up (48 cm^3).
The volume of carbon dioxide formed = (164 − 116) cm^3 = 48 cm^3 (this was absorbed by the sodium hydroxide solution).
The remaining gas was excess oxygen, 116 cm^3
Volume of oxygen *used* = (212 − 116) cm^3 = 96 cm^3
∴ 48 cm^3 methane react with 96 cm^3 oxygen to form 48 cm^3 of carbon dioxide.
1 volume methane reacts with 2 volumes of oxygen to form 1 volume of carbon dioxide.
The ratio of reacting volumes is 1:2:1, which is in accordance with Gay Lussac's law.

Avogadro's law, vapour density and molar volume
Avogadro's law states that equal volumes of different gases (measured at the same temperature and pressure) contain an equal number of molecules.

*The **vapour density** of a gas is the mass of a certain volume of the gas compared with the mass of an equal volume of hydrogen, at the same temperature and pressure.*

Avogadro's law can be used to derive the expression:

relative molecular mass = 2 × vapour density

Some syllabuses may require you to know this relationship. You could be given the percentage composition by mass of a gas and also its vapour density, from which you could calculate its empirical formula (from the percentage composition), its molecular mass (twice its vapour density), and hence its molecular formula (p. 105).

Avogadro's law can also be used to show that *the volume of a mole of any ideal gas at stp occupies 22.4 dm^3*. This volume is often called the **molar volume**.

The molar volume at room temperature and pressure (rtp) is approximately 24 dm^3.

Mass and volume calculations involving gases

Calculations of this type may involve up to four different steps.

1. The first step in *all* calculations of this type is to produce a fully balanced equation for the reaction.
2. Use the equation and the masses of reactants to calculate the number of moles of gases being formed.
3. Convert the number of moles of gas into a volume at stp (or rtp) by using the fact that one mole of a gas occupies 22.4 dm^3 at stp (24 dm^3 at rtp).
4. If necessary, convert the volume at stp into the volume the gas would occupy at the conditions stated in the question.

The following example should make this clear.

Calculate the volume of carbon dioxide (measured at 25 °C and 790 mm pressure) evolved when 10.0 g of potassium hydrogencarbonate are completely decomposed by heating.

Step 1 $\quad 2KHCO_3(s) \rightarrow K_2CO_3(s) + CO_2(g) \quad\quad + H_2O(g)$

Step 2 $\quad$ 2 moles $\longrightarrow$ 1 mole
$\quad\quad\quad\;\;$ 200 g $\longrightarrow$ 1 mole
$\quad\quad\quad\;\;$ 10 g $\longrightarrow \dfrac{10}{200} \times 1$ mole $= \dfrac{1}{20}$ mole

Step 3 $\quad \dfrac{1}{20}$ mole of carbon dioxide at stp occupies $\dfrac{1}{20} \times 22.4$ dm^3

Step 4 $\quad \dfrac{P_1V_1}{T_1} = \dfrac{P_2V_2}{T_2}, \quad \therefore V_2 = \dfrac{P_1V_1T_2}{T_1P_2}$

$$\therefore V_2 = \dfrac{760 \times \dfrac{22.4}{20} \times 298}{273 \times 790} \text{ litres} = 1.17 \text{ } dm^3$$

Questions

1 An electric current of 0.1 A is passed through an electrolyte for 1.25 hours. The quantity of electricity used (in coulombs) is: (a) 0.125; (b) 0.1; (c) 125; (d) 450; (e) 4500.

2 How many faradays are needed to produce: (a) 2.70 g of aluminium; (b) 6.0 g of magnesium; (c) 10 g of hydrogen; (d) 71 g of chlorine.

3 Two voltameters were connected in series with a battery. The first voltameter had platinum electrodes in dilute sulphuric acid and the second had copper electrodes in copper(II) sulphate solution. After one hour the cathode of the second voltameter had increased its mass by 0.3175 g.

Give ionic equations for the reaction at: (a) the cathode of the second voltameter; (b) the anode of the first voltameter.

Calculate: (c) the mass of the gas evolved at the anode of the first voltameter; (d) the volume occupied by this gas at s.t.p. (JMB)

4 State: (a) Gay-Lussac's law of gaseous volumes; (b) Avogadro's law.

Describe any one experiment you choose which fully illustrates Gay-Lussac's law.

Four grammes of sulphur on heating in oxygen give 2.8 litres of sulphur dioxide. Calculate the molecular mass of sulphur dioxide. (AEB)

5 State Gay-Lussac's law of combining volumes.

200 cm^3 of a gaseous element X_2 reacted with 650 cm^2 of a gaseous element Y_2 to form 450 cm^3 of a mixture of XY_3 and Y_2. It was later found that 50 cm^3 of excess of Y_2 remained unused. All volumes were measured under the same conditions of temperature and pressure.

(a) What volume of XY_3 was formed in the reaction?

(b) Write a statement to show the relationship between the volumes of X_2 and Y_2 used and the volume of XY_3 formed.

(c) Give a balanced molecular equation for the reaction. (JMB)

17 Preparing for the examination

17.1 Revising for the examination

Revising is very much an individual thing, and techniques which may be suitable for some students may be quite unsuitable for others. Nevertheless, there are certain points which are important for everyone.

It is essential to **understand** the principles of chemistry before any attempt is made to **learn** them. Read all of your notes in conjunction with this book and try to understand the principles you have been taught and also the point of each of the experiments you have done. If you do this several times you will find that the ideas become clearer, and this is made easier if there is no pressure on you to learn the work at the same time. You will almost certainly see how some of the ideas depend on each other, and how others interconnect, whereas if you try to learn the work as a small package at a time you will fail to see these connections.

Attempts to **learn** the facts and principles should only be made after you fully understand them; it is exceedingly difficult to learn something which you do not understand. Facts which are difficult to remember, and exceptions to general rules, should be amongst the last items you try to learn, and a condensed version of these should be revised the night before the examination. If you have planned your revision properly, there should be no need to revise **ideas** just before the examination.

Modern examination questions often expect you to be able to link together ideas which have been taught separately. For example, you might think of 'bonding' as the topic in which you learned to draw bonding diagrams for ionically bonded and covalently bonded compounds. Bonding is much more than this, however, for you have used the idea to explain why electrolytes have to be ionic, how giant ionic or giant atomic structures are so different from molecular

Preparing for the examination 219

structures, and why compounds have different properties from their constituent elements. It is very useful to vary your revision programme by thinking of one of the basic principles of chemistry, e.g. bonding, and seeing how this particular idea is used in different topics. Repeat this with atomic structure (leading to the activity series, changes across a period and down a group in the Periodic Table, equilibrium, strong and weak acids and many industrial processes) and others of your own choice. This varied approach will ensure an even better understanding of the principles, you will see connections you have not seen before, you will realize that some topics simply use the same idea in a different disguise, and you will begin to see the subject as a whole.

It is also important to understand that a *teaching* programme can sometimes result in detailed experimental accounts (e.g. in your notebook) which are themselves unlikely examination material. They were essential at the time as a means to an end, but what matters in the revision programme is that you learn the facts and principles shown by the experiment rather than the 'apparatus and method', etc. which were used. You will have noticed that many Units in this book make little or no reference to particular experiments. This is because it is most unlikely that you will be asked to describe an experiment which shows, for example, that some metal nitrates decompose on heating to form the oxide, water and nitrogen dioxide. You may have written up an experiment which shows this, but it is more important that you learn only the results of such experiments, as facts. Obviously there are some areas where you could be asked to describe experiments in full detail, but these tend to be general *techniques*, e.g. a distillation, chromatography, a salt preparation, a melting point determination and one to prove how reaction rates depend upon temperature. The more likely candidates for such questions have been indicated at the appropriate points in the book.

A vital part of any revision programme is the familiarization with the types of questions which can be expected, an understanding of what the examination papers are like, and the recommended times to be spent on each section. Above all, it is important to practise answering questions from previous papers, and to make sure that your timing of such answers is suitable for the examination itself. The number of papers which are set, and the individual questions within them, vary in length and style from one Examining Board to another, but your school or college should be able to show you some typical papers. Alternatively, papers are available for a reasonable charge from the individual Examining Boards (addresses in Preface).

Detailed advice on answering individual questions is given in the next few sections, which also contain specimen answers to typical questions.

17.2 Answering multiple choice questions

Most chemistry examinations include some multiple choice questions, either in a separate section or in a separate paper. There are several variations, but in all cases each question is accompanied by five alternative answers (lettered A to E), only one of which is correct. You have to decide which one of the five alternatives best answers each question.

You should remember the following points when answering multiple choice questions.

1 Each question scores exactly the same mark, but some are easier than others. When you come across a question which is obviously going to take a while to answer (e.g. one involving several mathematical manipulations) then leave it out on your first run through the questions. Too many students run out of time by battling with difficult questions when some of the questions they have not attempted may be comparatively simple. Your first run through the questions should therefore be an attempt to pick up the easy marks quickly by ignoring any questions you find difficult. Then attempt to answer as many of the remaining questions as possible in the recommended time for the section.

2 There is no excuse for failing to *attempt* every single question in the multiple choice part (on your final run through the section), because you should be able to make at least an 'intelligent guess' at the questions you find very difficult. This is a perfectly legitimate technique, and it means that you are using your knowledge of chemistry to eliminate *some* of the answers, even if you cannot decide which answer is the correct one. If you eliminate three answers, and simply cannot decide which of the remaining two is correct, then guess. You have used your knowledge to reduce the odds, and it is far better to do this than to ignore the question altogether or to guess one answer from five. Remember that marks are not deducted for incorrect answers.

3 It is advisable to read all five answers, even if you think that one of the early ones (e.g. A or B) is the correct answer. It is so easy to misread a question or to choose a partly correct answer, especially under examination conditions. You may at first choose answer A

Preparing for the examination 221

for a question but when you later read, say, answer D, you may realize that D is a better answer than A. Equally, it may be that a later answer, although itself not the correct one, triggers off a different way of thinking which enables you to see that your first choice is incorrect.

4 In those questions which involve calculations, it is often unnecessary to complete the calculation accurately. Remember that the accurate calculation has been done for you; you are provided with the exact answer and all you have to do is to recognize it. Suppose that the alternative answers to a question are 5, 15, 30, 40 and 60. You may be able to make a rough calculation in seconds which shows that the answer is *about* 55. A quick glance at the alternatives shows that the *exact* answer must therefore be 60, which is nearer to 55 than 40 is. You might waste precious time trying to produce an exact answer by calculation.

17.3 Answering structured questions

Many chemistry examinations include a section of structured questions, in which questions are broken down into a number of parts, each of which requires only a short answer. You are normally required to answer each part in the space provided on the question paper, and the marks allocated for each part are often indicated alongside the question.

The mark allocation and the space provided for each answer is an indication of the depth of answer required. Most students should be able to make an adequate answer in that space, and you must remember that no amount of extra writing (even on a topic you know a lot about) will gain you any more marks than the number allocated to that part of the question. If you find difficulty in expressing yourself in the space available, or if you are running short of time, it is acceptable to use phrases rather than sentences. Where time is desperately short it may be possible to score some marks by simply writing a list of words separated by dashes, provided that the words are in the correct order. The following example should make this clear.

Suppose that you are asked to describe briefly how you would make a sample of β-sulphur, and there are only four lines in which to answer. Pause and think before you start to answer, because you could easily use up 15 lines in making a description of the preparation. The problem is deciding what are the *important* steps to indicate. You could write in sentences, e.g.: 'Melt some roll sulphur gently in a test-tube, and pour it carefully into a cone made from filter paper. When a crust of sulphur forms over the surface, break open the cone to expose the crystals of β-sulphur.'

It is just possible to write the above answer in four lines, and it contains the essential points. Compare this with an attempt such as the following: 'I would take about 2 g of powdered roll sulphur and place it in a test-tube. I would heat the test-tube with a Bunsen burner until the sulphur melted to a pale yellow liquid. I would then pour the liquid into a paper cone made by folding several pieces of filter paper together. As the sulphur cools, a crust of solid sulphur begins to form over the surface of the liquid, and if the paper is then broken open long, needle-shaped crystals of β-sulphur can be seen hanging from the underside of the crust.'

You might be tempted to write an answer like this (if you have done your revision properly and if a continuation sheet is available for your answer) but such a detailed response would be more appropriate for a longer question, as in 17.4. As an answer to a structured question it would gain no more marks than the four lines given earlier, if only four lines are provided for your answer. You must be ruthless in deciding what the question is asking for; a brief description does not usually demand colours, appearance of crystals or a description of routine events like using a Bunsen.

If you find it difficult to write in a condensed style, or if time is short, try a series of short statements, e.g. 'Melt some roll sulphur – pour into paper cone – allow to form solid crust – break open cone – β-sulphur crystals will be hanging from underneath the crust.'

This technique can also enable you to provide a little more detail, if relevant, in the space available. When students are running short of time in an examination, they often fail to realize that they could cover the remaining written questions reasonably adequately by adopting this style. Remember that the first few marks available for an answer are always the easiest to obtain; it is most unwise to answer some questions in great detail and to leave others unanswered, for you could easily pick up valuable marks by answering the last question(s) in brief form. This point applies equally to the longer type of answers.

17.4 Answering longer questions

These are questions which normally take between 15 and 30 minutes to answer, according to the examination syllabus being followed.

Preparing for the examination 223

They can take many different forms, such as an essay, a detailed experimental account or a question broken down into several smaller parts.

Most sections involving longer questions offer you a choice. Make sure that you choose carefully. Inevitably this takes time, and you must attempt to think calmly about each question before you make a decision. Only too often students read part of a question quickly and then immediately make a decision to do it or to leave it, only to find too late that it is not really what they thought it was. Do not be put off by long, complicated-looking questions or those which at first look unfamiliar. These are often the easiest to answer if only you are prepared to think about them for a few minutes. (See how short the actual written answer is for the specimen answer given on p. 227.) Time spent on careful consideration of all of the questions is time well spent, and you must not panic into making hasty decisions.

Make sure that you spend approximately the recommended time on each of the questions that you choose. Questions which involve calculations or working out the names of 'unknown' chemicals from information provided can often be answered in less time than descriptive questions. If this happens, do not spend more than a moment or two wondering whether you have missed something out, but do spend the extra time so gained in a sensible way. For instance, you may be pleasantly surprised how easy it is to write the actual answer if only you take time in planning it first.

Ensure that you attempt the required number of questions. If you are supposed to choose three, remember that it is far, far better to produce $2\frac{1}{2}$ reasonably good answers than to spend all your time trying to produce 2 perfect answers on questions you think you can do well.

As the types of question vary enormously, other examination tips and techniques are best illustrated by a more detailed consideration of a few typical questions in the remainder of this Unit. Specimen answers are provided for these questions, and an analysis is given to indicate the kind of thinking required in planning the answer. As space is at a premium in this book, it is not possible to include many detailed analyses of questions, but some further tips on certain types of question are given at the end of the Unit.

A question involving oxidation and reduction

Give three definitions of the term reduction. Read the following descriptions of reactions and state in each case which of the substances give rise to the observation printed in italics. State also whether each product is formed by oxidation or reduction.

(a) When hydrogen sulphide is passed through a *yellow solution* of iron(III) chloride, a *yellow precipitate* is formed, suspended in a *pale green solution*. (b) When carbon monoxide is passed over hot lead(II) oxide, a *silver coloured liquid* is formed, and a *colourless gas*. Give an equation for the reaction. (c) If chlorine gas is passed through potassium iodide solution, a *dark coloured liquid* is formed. Give an ionic equation for the reaction.

Analysing the question

The first part simply asks for three definitions, which you should know. Before you can answer the remaining parts, you must be absolutely clear about what is being asked. Read the instructions several times if necessary. In each reaction (a, b, c) we are told that there is oxidation and reduction, so we have to recognize first what the products are and how they have been formed.

In (a) the chemicals have the formulae H_2S, $FeCl_3$ and H_2O (always remember that water is present in aqueous solutions). The only yellow precipitate which could be formed from these atoms is sulphur. The hydrogen sulphide thus becomes sulphur by losing hydrogen, i.e. it is oxidized. The iron(III) compound will thus be reduced, and this will produce an iron(II) salt. Iron(III) salts are typically yellow in solution, and iron(II) salts are green. Note that these deductions have not been reached by trying to recall a particular reaction involving hydrogen sulphide and iron(III) chloride, but it may be that you find it easier to reach the same conclusion by remembering this particular reaction or by remembering the fact that hydrogen sulphide is a reducing agent.

In (b), the only silvery substance which could be formed from the atoms present is metallic lead, in which case the carbon monoxide is removing oxygen from the lead(II) oxide (reducing it), so becoming the colourless gas carbon dioxide.

In (c) the only dark solid which could be formed from the atoms involved is iodine; this is a displacement reaction and also an oxidation (see p. 161).

A typical answer to this question

(i) Reduction is a process in which oxygen is removed from a compound. (ii) Reduction is a process in which hydrogen is added to a substance. (iii) Reduction is a process in which a substance gains one or more electrons.

(a) The yellow colour of the solution is caused by the iron(III) ions, the pale green colour by the iron(II) ions, and the yellow precipitate is sulphur. In the reaction the hydrogen sulphide is oxidized to sulphur by the loss of hydrogen, and iron(III) ions are reduced to iron(II) ions by gaining electrons.

(b) The silver coloured substance is metallic lead and the colourless gas is carbon dioxide. The lead(II) oxide is reduced to lead by loss of oxygen, and the carbon monoxide is oxidized to carbon dioxide by gaining oxygen.

$$PbO(s) + CO(g) \to Pb(s) + CO_2(g)$$

(c) The dark solid is iodine. Iodide ions are oxidized to iodine atoms by losing electrons, and chlorine atoms are reduced to chloride ions by gaining electrons.

$$Cl_2(aq) + 2I^-(aq) \to I_2(s) + 2Cl^-(aq)$$

Notes about the answer

1 Notice how important it is to think about the question before answering it; this is not wasted time, because the actual written answer does not take very long.
2 Check the question again and make sure that every part has been answered fully. Notice that we were asked to account for each product but that they are not always mentioned in the question. In (c) for example, chloride ions are produced but they are not referred to in the question.
3 In every description of an oxidation or reduction, you must always state the substance being oxidized or reduced, *and* the substance it becomes, *and* why it is being oxidized or reduced.
4 Note that the different parts of the same question often have a close connection, and it is useful to look for this. Here there is an example of each type of reduction.
5 Always try to give an equation for any reaction you describe, even if not asked to do so. Note the request for an *ionic* equation in (c).
6 Note that the answers have been restricted to the particular questions being asked; the temptation to add any further detail has been avoided because there are no extra marks to be gained.
7 Another common type of question involving oxidation and reduction requires the name of an appropriate oxidizing or reducing agent which will bring about a named change. For this we need to recall particular situations we have studied. If these do not readily come to mind, make lists in rough of the common

oxidizing and reducing agents (p. 63). If a chemical is to be oxidized, try to think of a reaction between it and each of the oxidizing agents in turn until you recall a reaction which works. Repeat this with reductions.

Questions involving the identification of substances

In the following descriptions, the names of chemicals have been replaced by letters. Identify each lettered chemical, and give an equation for each of the reactions.

(a) When a white solid X was warmed with sodium hyrdoxide solution, gas Y was produced. The gas changed damp red litmus blue. When barium chloride solution was added to a solution of X, a white precipitate Z was formed. (b) When cold water was added to a grey metal A, hydrogen was produced and a white solid B, which was suspended in the solution. The suspension was filtered, and carbon dioxide was passed through the filtrate, when a different white solid C was formed. (c) A red powder D was warmed with dilute nitric acid, when a brown solid E was formed. E was then warmed with concentrated hydrochloric acid, and a gas F was given off which bleached moist litmus paper.

Analysing the question

In each part, look for a specific reaction which tells you *exactly* what something is. It is then comparatively simple to work forwards or backwards from this identified chemical in order to identify the others.

In (*a*) we should recognize two specific reactions; the gas Y *must* be ammonia as it is the only alkaline gas encountered at O level, and Z *must* be barium sulphate as this stage is the test for a sulphate (p. 170). X is therefore a sulphate, and as it gives off ammonia when heated with an alkali (sodium hydroxide) it must be an ammonium salt (p. 170), namely ammonium sulphate.

In (*b*) the only common metals which liberate hydrogen from cold water are calcium, sodium, lithium and potassium, and they form their hydroxides in doing so. The hydroxides of lithium, sodium and potassium are soluble in water (all sodium and potassium compounds are soluble) but calcium hydroxide is only partly soluble, so the metal is calcium and the first white suspension is calcium hydroxide. The filtrate is a dilute solution of calcium hydroxide (limewater) and this is the reagent used in the standard test for carbon dioxide, so that the second suspension is calcium carbonate (i.e. the milkiness in limewater when testing for carbon dioxide).

In (c) the only red powder studied in elementary chemistry is dilead(II) lead(IV) oxide (red lead oxide), Pb_3O_4 and the gas F which bleaches damp litmus paper will be chlorine. Recognition of E depends upon your recall of this particular reaction of the oxides of lead (p. 116), i.e. it is lead(IV) oxide, PbO_2. Alternatively, you may simply recall the fact that the only brown compound of lead is lead(IV) oxide.

A typical answer to this question

(a) X is ammonium sulphate and Y is ammonia.

$$(NH_4)_2SO_4(s) + 2NaOH(aq) \rightarrow 2NH_3(g) + Na_2SO_4(aq) + 2H_2O(l)$$

Z is barium sulphate.

$$Ba^{2+}(aq) + SO_4^{2-}(aq) \rightarrow BaSO_4(s)$$

(The full equation could be given instead of the ionic one.)

(b) A is calcium metal and B is calcium hydroxide.

$$Ca(s) + 2H_2O(l) \rightarrow Ca(OH)_2(s) + H_2(g)$$

C is calcium carbonate.

$$Ca(OH)_2(aq) + CO_2(g) \rightarrow CaCO_3(s) + H_2O(l)$$

(c) D is red lead oxide and E is lead(IV) oxide.

$$Pb_3O_4(s) + 4HNO_3(aq) \rightarrow 2Pb(NO_3)_2(aq) + PbO_2(s) + 2H_2O(l)$$

F is chlorine.

$$PbO_2(s) + 4HCl(aq) \rightarrow PbCl_2(s) + 2H_2O(l) + Cl_2(g)$$

Notes about the answer

1 Do not be put off from answering a question of this type if you are unable to answer just a small part of it. You can easily obtain maximum marks for each part you can answer, and yet you are having to do very little writing to gain them. The alternative might be a question involving a long written response, in which it is much easier to lose marks by leaving out details.

2 Notice how short the written response is, for questions of this type. Explanations are *not* given in the answer because the question only asked for the *name* of each labelled chemical and an *equation* for each reaction. It is assumed that you have made the necessary 'explanations', either in your head or in rough, in order to reach the answer.

3 You have ample time in questions of this type to work out a balanced equation providing that you know the names of the reactants and products. Do not worry, therefore, if you cannot actually remember the equation.

Some specific terms used in examination questions

(a) '*add slowly drop by drop*' or '*add until in excess*'. If a phrase of this type occurs in an examination question, always look for the possibility of *two* stages in the reaction, e.g. when sodium hydroxide solution is added to some metal salt solutions (p. 117). The average student only thinks of a one-stage reaction.

(b) '*Describe what you would see if . . .*' Descriptions of this type are often badly answered. You must start with the reactants, giving their names, their state (if solid say whether powder or crystals), colours and any other visible feature. Then do the same for the products, thus emphasizing the changes observed during the reaction. Note that a description of the reactants is relevant; without it you would not draw attention to the observed *changes*. The following examples show a good answer and a typically poor answer to the question 'describe what you would see when sulphur burns in air'.

A good answer: The yellow powder (sulphur) first melts when heated to form an amber coloured liquid. This quickly catches fire and burns with a blue flame to form the colourless gas sulphur dioxide (which is sometimes visible as a steamy gas because it fumes in moist air).

A poor answer: Sulphur burns to form sulphur dioxide.

Compare the two; note how many *observations* are recorded in the first answer (yellow, powder, melt, amber liquid, burns, blue flame, colourless gas, etc.) As the question specifically asked for what you would *see*, all of these points are relevant.

(c) '*under what conditions . . .*' Pupils rarely give a complete account of reaction conditions when asked to do so. For example, if this is relevant to the question, state whether the reaction takes place in solution, in the gas state or between pure liquids. If acids are involved, are they used in *dilute* or *concentrated* solutions? Is heating or a catalyst required? Can the reaction be carried out in the open laboratory or is a fume cupboard needed? Is a pressure higher than normal required?

(d) Some words used in examination questions have highly specific meanings, and it is easy to answer a question incorrectly unless you recognize exactly what is intended. For example, the instruction 'state . . .' simply requires a factual description and *not* an explanation. This should be compared carefully with words such as 'describe' and 'explain' which would require a more detailed answer.

'Describe' could involve experimental conditions and/or observations, and 'explain' will obviously involve an explanation as to how or why something happened.

Answers to questions

UNIT 1
1. (d) 2. (d) 3. (e) 4. (c) 5. (d)
6. (a) WX_3; covalent (b) XZ; ionic (c) X_2; covalent (but not a compound) (d) No combination (e) Metallic lattice formed, not a compound.
9. (a) 17 (b) 17 (c) 20 (d) 18

UNIT 2
1. (c) 2. (b) 3. (b) 4. (i) (d) (ii) (e) (iii) (b) 8. (c)

UNIT 3
1. (b) 2. (a) 3. (c) 4. (a) False (b) True (c) False (d) True (e) False 11. (c)

UNIT 5
1. (c) 2. (e) 3. (c)

UNIT 6
1. (c) 2. (d) 3. (d) 4. (a) Metallic (b) A (c) E 6. (d) 7. (d)
8. (a) W and Y (b) V and Z (c) X

UNIT 7
1. (e) 2. (b)
3. (a) (i) 164 g (ii) 161 g (b) (i) 60 g (ii) 154 g (iii) 78 g (c) (i) 2.5 (ii) 0.2 (iii) 1.0 (iv) 0.25
4. (a) N_2O (b) Na_2O 5. $Zn\,CO_3$ 6. (a) CH_2Br (b) 188 (c) $C_2H_4Br_2$
7. HCN 8. (c) 64 (d) 0.56 litres at stp

UNIT 8
2. (d) 3. (e) 4. (d) 5. A is $CaCl_2$ B is $CuCO_3$

UNIT 9
5. (a) 6. (b) 9. (e)

UNIT 10
3. (d) 4. (c) 5. (a)

Answers to questions 231

UNIT 11
1. (c)
2. (a) Iron(II) sulphate (b) calcium chloride (c) manganese(IV) oxide (d) washing soda (e) iron(III) oxide
4. Formular mass = 56   Atomic mass = 39
5. 70 cm^3 of gas, consisting of 30 cm^3 excess oxygen and 40 cm^3 carbon dioxide
6. 0.05 moles carbon dioxide 1.12 litres at stp

UNIT 12
1. (c)   2. (d)   3. (d)   4. (e)
5. A is ethyne B is ethene D is 1,2-dibromoethane E is polythene

UNIT 13
No answers

UNIT 14
1 (a) 4 g (b) (i) more methanol (ii) more methanol (iii) more methanol
5. (a) (i) more ammonia (ii) more ammonia (b) evolved
6. (a) more sulphur trioxide (b) less sulphur trioxide (c) more sulphur trioxide
7. (b) and (e)

UNIT 15
2. (c)

UNIT 16
1. (d) 2. (a) 0.3 (b) 0.5 (c) 10 (d) 2
3. (a) $Cu^{24}(eq) + 2e \rightarrow Cu(s)$ (b) $4OH^-(eq) - 42 \rightarrow 2H_2O(l) + O_2(g)$
(c) 0.08 g oxygen (d) 0.056 litres   4. 32
5. (a) 400 cm^3 (b) 1 vol. X_2 + 3 vols. Y_2 → 2 vols. XY_3 (c) $X_2 + 3Y_2 \rightarrow 2XY_3$

Periodic Classification

The table gives the symbols, atomic numbers, and re

Group	I	II					Transition
Period 1	1 H 1·008						
Period 2	3 Li 6·941	4 Be 9·012					
Period 3	11 Na 22·99	12 Mg 24·31					
Period 4	19 K 39·10	20 Ca 40·08	21 Sc	22 Ti	23 V	24 Cr 52·00	25 Mn 54·94
Period 5	37 Rb	38 Sr 87·62	39 Y	40 Zr	41 Nb	42 Mo	43 Tc
Period 6	55 Cs	56 Ba 137·3	57* La	72 Hf	73 Ta	74 W	75 Re
Period 7	87 Fr	88 Ra	89 Ac	90 Th	91 Pa	92 U 238·0	93 Np

*The 14 'lanthanons',

Elements (See Section 5)

masses (four significant figures) of the commoner elements.

			III	IV	V	VI	VII	VIII or 0
								2 He 4·003
			5 B 10·81	6 C 12·01	7 N 14·01	8 O 16·00	9 F 19·00	10 Ne 20·18
			13 Al 26·98	14 Si 28·09	15 P 30·97	16 S 32·06	17 Cl 35·45	18 Ar 39·95
28 Ni 58·70	29 Cu 63·55	30 Zn 65·38	31 Ga 69·72	32 Ge 72·59	33 As 74·92	34 Se 78·96	35 Br 79·90	36 Kr 83·80
46 Pd 106·4	47 Ag 107·9	48 Cd	49 In	50 Sn 118·7	51 Sb 121·8	52 Te	53 I 126·9	54 Xe 131·3
78 Pt 195·1	79 Au 197·0	80 Hg 200·6	81 Tl	82 Pb 207·2	83 Bi 209·0	84 Po	85 At	86 Rn
96 Cm	97 Bk	98 Cf	99 Es	100 Fm	101 Md	102 No	103 Lr	

clusive, are omitted here.

Index

absolute zero 14
acidic solutions 68
acids 67
  definition of 68
  properties of 67
  strength of 76
activity series 58
  displacement reactions 59
addition polymers 182
addition reaction 175
air, liquefication of 210
  pollution of 81
  reaction with metals 58
  solution in water 79
air pressure, effect on boiling
    point 16
alcohols 178
alkali 69
alkenes 174
allotropes 95
  of carbon 95
  of sulphur 98
aluminium 127
  extraction of 52
  reaction of metal with oxygen 58
  reaction with dilute acid 60
  reaction with water or steam 58
aluminium chloride 124
aluminum compounds 128
aluminium hydroxide 117
aluminium nitrate 120
amino acids, proteins from 183
ammonia, chemical properties 156
  Haber process 205
  physical properties 156
  prepartion of 156
  shape of molecule 32
  test for 169

ammonia – soda process 209
ammonia solution, preparation of 156
  properties of 157
  uses of 157
ammonium carbonate 157
ammonium chloride 157
ammonium compounds, action with
    bases 70
  test for 170
  uses of 157
ammonium nitrate 157
ammonium salts 157
ammonium sulphate 157
amphoteric oxides 85
anhydride, of acid 85
anion, definition of 49
  in electrolysis 50
anode 49
atom, definition of 9
  size of 10, 26
atomic mass 13
atomic number 11
atomic structure 11
Avogadro's constant (number)
  100
Avogadro's law 215

barium, flame test 170
barium sulphate 122
base, definition of 69
  properties of 70
battery (dry cell) 57
bauxite 52
bleaching powder 127
boiling point 16
  determination of 46
  effect of impurities on 47
  effect of pressure changes on 47

Index

bonding 23
Boyle's law 18
Bronsted–Lowry theory 68
Brown ring test 121
Brownian movement 19
butane 173

calcium, compounds of 126
  flame test 170
  properties of 126
  reaction of metal with oxygen 58
  reaction with dilute acid 60
  reaction with water or steam 58
  uses of 127
calcium carbide 126
calcium carbonate 119
  uses of 127
calcium nitrate 120
calcium oxide, preparation of 115
  properties of 116
calcium phosphate 127
calcium sulphate 122
carbohydrates 184
carbon, allotropes of 95
  as a reducing agent 63
  chemical properties 145
  combustion in oxygen 84
  physical properties 145
carbonates, action with acids 68
  metallic carbonates,
    preparation of 118
    properties of 119
  table 119
  tests for 170
carbon cycle 148
carbon dioxide, chemical properties 146
  effect on water 85
  formula and state 85
  physical properties 146
  preparation of 146
  shape of molecule 33
  test for 169
  type of oxide 85
  uses of 147
carbon monoxide, as a reducing agent 63
  chemical properties 146
  effect on water 85
  physical properties 147
  type of oxide 85

cast iron 129
catalyst 198
cathode 49
cation, definition 49
  in electrolysis 50
change of state 14
Charles' law 17
chemical industry 205
chlorides, metallic chlorides 123
  preparation of 123
  properties of 123
  tests for 169
chlorine, as an oxidizing agent 63
  chemical properties 162
  isotopes of 13
  manufacture of 56
  physical properties 163
  preparation of 161
  test for 169
  uses of 163
chromatography 43
combustion, definition of 79
  heat of 189
compounds, definition of 37
  properties of 37
contact process 206
co-ordinate bonding 29
co-ordination number 95
copper, compounds of 131
  flame test 170
  reaction of metal with oxygen 58
  test for 170
  uses of 131
copper(II) carbonate 119
copper(I) chloride 124
copper(II) chloride 124
copper(II) hydroxide 117
copper(II) nitrate 120
copper(II) oxide, preparation of 115
  properties of 116
copper(II) sulphate 122
  electrolysis of 53
covalent bonding 26
covalent compounds 30
covalent molecules 32
cryolite 54
crystallization 39

dative bonding 29
deliquescence 134
detergents 134

236 Index

detergents – contd.
 how they work 137
 lowering surface tension 138
 removing grease and dirt 139
 soapy detergents 137
 synthetic detergents 137
Devarda's alloy 121
diamond 95
diaphragm cell 56
diffusion 18
 factors affecting rate of 19
displacement reactions 59
distillation, definition of 39
 of crude oil 41
Down's cell 54
dry cell 57

efflorescence 134
electrochemical series 56
electrode 49
electrode potentials 56
electrolysis, definition of 49
 examples of 52
 Faraday's laws of 212
 industrial uses of 52
 oxidation and reduction 62
 part water plays in 51
 what happens during 50
electrolyte 49
electron 11
electroplating 55
element 36
empirical formula 105
endothermic reactions 187
energy changes 187
equations, constructing and balancing 107
 determining reacting masses from 108
 determining reacting volumes from 109
 interpretation of 108
 ionic 107
equilibria 199
esterification 181
ethane 172
ethanoic acid 180
ethanol, chemical properties of 180
 laboratory preparation of 179
 manufacture and uses of 179
ethene, chemical properties 175
 laboratory preparation 175

 physical properties 175
 shape of molecule 33
 tests for 177
exothermic reactions 187

Faraday's laws of electrolysis 212
fertilizers 160
filtrate 39
flame tests 170
formula mass 13, 100
formula unit 101
fractional crystallization 137
fractional distillation 42
fractionating column 42
fractions 41
fuels 191
functional group 178

gases, mass and volume calculations 216
 preparation of 82
 volume changes in 214
gas pressure 17
gas state 14
Gay–Lussac's law 214
glass 149
Graham's law of diffusion 20
graphite 95

Haber process 205
halogens 161
hardness of water, permanent hardness 141
 temporary hardness 141
heat of combustion 189
heat of neutralization 190
heat of precipitation 189
heat of reaction 188
heat of solution 190
hexane 172
homologous series 172
hydration 133
hydrocarbons 172
hydrochloric acid, preparation of 164
 properties of 164
hydrogen, as a reducing agent 62, 63
 chemical properties 86
 physical properties 86
 preparation of 86
 test for 169
 uses of 87

Index 237

hydrogencarbonates 170
hydrogen chloride, chemical properties 164
 physical properties 164
 preparation of 163
 test for 169
hydrogen peroxide, as a bleaching agent 143
 as an oxidizing agent 63, 143
 decomposition 141
 physical properties 141
hydrogen sulphide as a reducing agent 63
 reaction with chlorine 162
 test for 169
hydrolysis 133
hydronium ion 30
hydroxides, preparation of metallic hydroxides 117
 properties of metallic hydroxides 117
hygroscopic 134

immiscible liquids 43
indicators, colour changes of 70
 universal 71
industrial processes 205
intermolecular forces 30
intramolecular bonds 30
ion, definition of 9
 in electrolysis 49
 positive ion 93
ionic bonding 23
ionic compounds 24
ionic radius 93
iron, compounds of 129
 conversion to steel 208
 oxidation and reduction 129
 production in blast furnace 206
 reaction of metal with oxygen 58
 reaction with dilute acid 61
 reaction with water or steam 58
 rusting of 128
 test for 171
 types of 129
iron(II) carbonate 119
iron(II) chloride 124
iron(III) chloride 124
iron(II) diiron(III) oxide 115, 116
iron(II) hydroxide 117
iron(III) hydroxide 117
iron(II) nitrate 120

iron(III) nitrate 120
iron(II) oxide 115, 116
iron(III) oxide 115, 116
iron(II) sulphate 122
isomerism 172
isotopes 12

kinetic energy 14, 16, 17
kinetic theory 14

latent heat of vaporization 188
law of conservation of mass 103
law of constant composition 103
law of multiple proportions 103
lead compounds of 129
 reaction of metal with oxygen 59
 reaction with dilute acid 60
 test for 171
lead(II) carbonate 119
lead(II) chloride 124
lead(II) hydroxide 117
lead(II) nitrate 120
lead(II) oxide 116
lead(IV) oxide 116
lead(II) sulphate 122
Le Chatelier's principle 200

magnesium, compounds of 126
 reaction of metal with oxygen 58
 reaction with dilute acid 60
 reaction with water or steam 58
 uses of compounds 127
magnesium carbonate 119
magnesium chloride 124
magnesium hydroxide 117
 use of 127
magnesium nitrate 120
magnesium oxide 115, 116
magnesium sulphate 122
 uses of 127
mass number 11
melting point, determination of 45
 effect of impurities on 46
mercury 59
mercury cell 56
metallic carbonates, preparation of 118
 properties of 119
metallic chlorides, preparation of 123
 properties of 123

Index

metallic hydroxides, preparation of 117
  properties of 117
metallic nitrates, preparation of 120
  properties of, 120
metallic sulphates, preparation of 121
  properties of 122
metals, action of dilute acids on 60
  as reducing agents 63
  displacement reactions 59
  extraction of 61
  metallic structures 94
  oxides and hydroxides as bases 69
  oxides, formulae and properties of 114
  properties of 36
  reaction with oxygen in air 58
  reaction with water or steam 58
methane, chemical properties 173
  formula 172
  physical properties 173
  shape of molecule 32
  tests for 177
mixture, definition of 37
  properties of 37
molar solution 101
molar volume 215
mole 100
molecular formula 106
molecular mass 100
molecular structures 98
molecule, definition of 9
  size of 10

neutral oxides 85
neutralization 71, 190
neutron 11
nitrates, heating of metallic nitrates 121
  preparation of metallic nitrates 120
  properties of metallic nitrates 120
  solubility of 121
  tests for 170
nitric acid, as an oxidizing agent 63, 159
  chemical properties 158
  laboratory preparation 158
  physical properties 158
  production of 210
  uses of 159
nitrogen, chemical properties 154
  in fertilizers 160

  oxides of 155
  physical properties 154
  preparation of 154
  test for 169
  uses of 155
nitrogen cycle 159
nitrogen dioxide, chemical properties 155
  effect on water 85
  formula and state 85
  physical properties 155
  type of oxide 85
non-metals 36
NPK values 160
nylon 185

octane 173
oil industry 177
organic acids 180
  reaction with alcohols 180

pentane 173
periodic table 89
  changes across a period 91
  changes down a metallic group 93
  changes down a non-metal group 93
  reactivity of metals 93
  reactivity of non-metals 93
perspex 182, 183
pH scale 71
phosphorus 84, 160
phosphorus(V) oxide, effect on water 85
  formula 85
  type of oxide 85
photosynthesis 80
pollution, of air 81
  of water 134
polymerization, addition 181
  condensation 181
polymers, natural condensation polymers 183
  synthetic addition polymers 182
  thermosetting 182
polystyrene 182, 183
polythene 182, 183
polyvinyl chloride 182, 183
potassium, compounds of 125
  flame test for 170
  in fertilizers 160
  properties of 60, 124

Index 239

potassium dichromate(VI) 63
potassium hydroxide 117
potassium manganate(VII) 63
potassium oxide 115
precipitation 189
propane 172
proteins 183
proton 11
purification methods 45

radical 21
rates of reaction 196
reactivity, of metals 93
  of non-metals 93
red phosphorus, combustion in
    oxygen 84
reducing agent, definition of 62
  examples of 63
reduction, definition of 62
  of iron(III) compound to iron(II)
    compound 130
relative molecular mass 13
residue 39
repiration 80
reversible reactions 199
rusting 128

salts, definition of 72
  preparation of 73
saponification 181
silica glass 149
silicon 148
silicon compounds 149
  glass manufacture 149
silicon dioxide, formula 85
  type of oxide 85
silver chloride 124
soda glass 124
sodium, compounds of 125
  extraction of 54
  flame test 170
  properties of 60, 124
  reaction of metal with oxygen 58
  reaction with water and steam 58
  uses of 125
sodium carbonate 119
  manufacture of 209
  uses of 125
sodium chloride 124
  electrolysis of 53
  uses of 125

sodium hydrogencarbonate 125
sodium hydroxide 117
  electrolysis of 53
  manufacture of 56
  uses of 125
sodium monoxide 116
sodium nitrate 120
sodium peroxide 116
sodium sulphate 122
solids, drying of 45
  melting point 46
  purity of 45
solubility curves 136
  uses of 136
solute 39
Solvay process 209
solvent 39
standard solution 167
starch 184
steel 129
structure, giant atomic lattices 95
  giant ionic structures 97
  giant structures 94
  metallic structures 94
  molecular structure 98
substitution reaction 173
sulphates, preparation of metallic
    sulphates 121
  properties of 122
  tests for 170
sulphur, chemical properties 150
  combustion in oxygen 84
  extraction 149
  physical properties 150
  uses of 150
sulphur dioxide, as a bleaching agent
    152
  as an oxidizing agent 152
  as a reducing agent 151
  chemical properties 151
  effect on water 85
  formula 85
  physical properties 151
  preparation of 151
  test for 169
  type of oxide 85
sulphuric acid, as an oxidizing agent 63
  concentrated sulphuric acid 153
  electrolysis of 53
  manufacture by Contact process 206
  uses of 154

sulphurous acid 154
sulphur trioxide, chemical properties 153
  effect on water 85
  formula 85
  physical properties 153
  preparation of 152
  type of oxide 85
symbols 21

terylene 185
thermoplastic 182
titrations 167
transition elements 92

unit cell 97
universal indicator 71

valencies 22
van der Waals' forces 31
vapour density 215
volumetric analysis 167

water, as a chemical 133
  formula and state 85
  hardness of 139
  methods of softening 141
  permanent hardness 141
  pollution 134
  reaction with metals 58
  shape of molecule 33
  temporary hardness 141
  type of oxide 85
water of crystallization 133
wrought iron 129

zinc, compounds of 126
  reaction of metal with oxygen 59
  reaction with water or steam 58
  reaction with dilute acids 60
  test for 170
  uses of 127
zinc carbonate 119
zinc chloride 124
zinc hydroxide 117
zinc nitrate 120
zinc oxide, preparation of 115
  properties of 116
zinc sulphate 122